HISTORY
THROUGH
THE EYES OF FAITH
Western Civilization
and the Kingdom of God

Ronald A. Wells

Christian College Coalition
For Enduring Values

HarperSanFrancisco
A Division of HarperCollinsPublishers

The Christian College Coalition is an association of Christian liberal arts colleges and universiti⁃s across North America. More than 30 Christian denominations, committed to a variety of theological traditions and perspectives, are represented by our member colleges. The views expressed in this volume are primarily those of the author(s) and are not intended to serve as a position statement of the Coalition membership.

Library of Congress Cataloging-in-Publication Data

Wells, Ronald A. 1941-
   History through the eyes of faith.

   1. Civilization, Occidental—History.
2. Civilization, Christian—History. I. Title.
CB245.W45   1989   909'.09821   88-45717
ISBN 0-06-069296-0

92 93 MAL 10 9 8 7 6 5 4

OTHER BOOKS IN THIS SERIES:

CONTENTS

FOREWORD

What does the history of the West look like when seen by a highly competent historian through the eyes of faith? Someone might be inclined to ask in reply, "Doesn't it look the same as it does to any other competent historian?" Reading this book will show you otherwise. Professor Wells discusses this very issue of the "objectivity" of history. For as he tells the story of the West, he stops every now and then to reflect on some of the questions that the craft of history as such, and his own telling of the story of the West, might raise in the mind of a reflective Christian.

Most people today would agree that history is not merely a matter of getting the facts right. Always it is a matter of interpreting facts in the face of the data available. Not surprisingly, competent scholars frequently differ in their interpretations. Professor Wells's telling of the story of the West is different from the story that non-Christians would tell. But it is also different from how various competent *Christian* historians would tell the story. Professor Wells recognizes this, and he takes a stand. But he acknowledges that the positions he takes do not make his telling of the story *the* Christian interpretation of Western history, only *a* Christian interpretation. It is one position in the dialogue. And we get an added bonus: Professor Wells not only takes a position in the ongoing debates among historians, both Christian and non-Christian, but often tells us what those debates are all about.

The underlying thesis of the book is that we in the West—like all humanity—have moved through a succession of crises. A society lives with a certain ideal vision of reality and the good, and it attempts to interpret its experience in the light of that vision and to guide its actions by it. But sometimes the ideal and the real move so far apart that attempts to bring them together are sensed, by society at large, to be failures. The society is then in crisis. Professor Wells's story of Western humanity is the story of the crises through which we have moved.

In this book our present situation is illuminated. But more than that: over and over readers will find themselves challenged to reflect on the light that the Christian gospel throws on history and the light that history throws on the Christian gospel. Christians with settled views on the Middle Ages, say, or on the Reformation, may well find themselves challenged to reconsider those views. And non-Christians with settled views on, say, the glories of the modern world, may well find themselves forced to reconsider those. What is important about the book is its passion for wholeness, for integrity. Professor Wells tries to get beyond the bits and pieces to see Western history whole. He struggles to bring together his Christian faith with the history he knows. In our culture, which, while longing for wholeness, finds almost everything piecemeal, that is remarkable.

Nicholas Wolterstorff
Professor of Philosophy
Calvin College, and the Free University of Amsterdam

ACKNOWLEDGMENTS

This book is the responsibility of one person but the result of the work of many persons. Throughout my writing I was helped and encouraged by many people. First among those who stood with me are the members of the task force of the Christian College Coalition: Nicholas Wolterstorff, Timothy Smith, Mark Noll, Russell Bishop, Shirley Mullen, James Juhnke, and James Cameron. Their insightful criticisms made this a better book. My colleagues in the Calvin College history department—especially Frank Roberts—were supportive and helpful, both in encouraging me and in criticizing the manuscript. James Bratt, David Diephouse, and William Van Vugt also gave me special helps at critical times, as did Del Ratzsch of Calvin's philosophy department. The provost of Calvin College, Gordon Van Harn, worked with the Christian College Coalition's John Dellenback and Karen Longman to provide logistical support, without which the project could not have been undertaken. I am very grateful to them and especially to Karen Longman, whose grace, patience, and humor were invaluable.

In June 1988, the Christian College Coalition staff, led by Jerry Herbert, Susan Baldauf, and Sandy Hoeks, organized a conference to discuss this book. Many of the conference participants offered valuable criticism that strengthened the book. Those I would particularly like to thank are Daniel Jensen, Steven Fratt, Steven Smiley, Joel Carpenter, Clyde Greer, David Wollman,

Richard Pierard, Paul Kubrict, James Hertzler, John Oliver, and Ted Davis.

There is one other person to be mentioned—now numbered among the saints "on another shore and in a greater light"— whose life and work gave me, and this book, its initial inspiration: Abraham Kuyper (1837–1920), prime minister of the Netherlands and founder of the Free University of Amsterdam. He is the hero of my intellectual pilgrimage because it was his work that first gave me to believe that I could go beyond being a Christian who is a scholar to being a Christian scholar.

This book will be a success if it helps student readers to take up the academic cultural task of "kingdom scholarship," in the conviction that our hearts and minds are united in the declaration of the lordship of Christ in all areas of life.

Ronald A. Wells

Chapter 1

AN INVITATION TO HISTORY:
A CHRISTIAN CALLING

This is a Christian book. By *Christian* I mean that a Christian wrote it primarily for Christians. Nonbelievers are welcome to listen to the discussion to follow, and they may even benefit from it. But, this work assumes that Christianity—both as personal faith and as worldview—is normatively correct. We will enter into no apologetics in this book, in that we will not try to defend or to explain the worth of Christian commitment. While that may be a valid exercise, it will not be done here.

This is a history book. History is the study of humans and time, indeed, of humans changing over time. Furthermore, history is the memory of the stories about people changing over a time span. In a certain sense, history would not be possible if it were not for the telling of it. Like Lord Berkeley's notional trees falling in the forest (if trees fell with no one to hear them, he asked, would there be any sound?), history untold is not history at all and, technically, may not even exist. History, therefore, is vital to our human existence. To have no story is, almost, to have no life. People suffering from amnesia can live and function, but they lead pitiable lives because they have lost contact with their own story. When societies and cultures lose contact with their own stories, they are also pitiable.

This is a Christian history book. Christians, in the view of this book, should have a considerable interest in history precisely because they are people of a story. While Christianity surely has a personal element, it is never, strictly speaking, personal. Despite their individuality, Christians find their true identity firmly rooted in a collectivity: We are not alone in this life but members one of another. The kingdom, as we say in the language of faith, has come, is present, and is yet to come. And our collective membership in that kingdom rests on a common affirmation of a story. Christians *are* Christians not solely because they made a "decision for Christ" but because they became "members incorporate" of Christ's Body. If anybody, then, should be interested in history, Christians should.

This is a Western Christian history book. *Western* denotes that civilization that is distinct from, say, those found in Africa, south Asia, and east Asia. I accept that even the terms are difficult (West of what? East of what?). But we use the term herein as men and women *in* Western civilization itself have used it. We will talk about that self-consciously different civilization that began in what we now call the "Middle East" (indeed, in the "middle" of what?), whose story encompasses Europe and the Americas (as well as those other outposts overseas where Westerners migrated). There may be some justification for saying that Christianity is an "Eastern" religion, like all the main religions of the world. But it has been associated with Western culture ever since the Jewish missionaries made it so successful among the Greeks and the Romans. Yet, and this should be an important clue to the argument of this book, while Christianity has long been *associated* with Western civilization, it would be wrong to *identify* it with Western civilization. Still, it is important for Christians to sort out the story of Western civilization because the readers of this book are, and the culture in which they live is, surely Western. A visit to Calcutta or Tokyo will remind them of this.

This is an honest Western Christian history book. *Honest* means more than merely telling the truth in factual terms but also telling the truth in all its ambiguity and complexity. Honest history differs from ideological history, in which the story comes "out right," according to the writer's values. While history is usable in understanding ourselves, if we approach history mainly to find a "usable past" with which to support an ideology or to advance a program, then we have not really studied history. There *are* some times when "our side" does the wrong thing and "their side" the right. Sometimes Christians embarrass us and non-Christians attract us. As Christians we "see through a glass darkly," and it does no good to deny that. Knowing the "author of truth" gives us an advantage in knowing truth over our secular neighbors, but it does not insure that we know the truth, which surely exists in the mind of God but comes ambiguously to us. Once in a while we experience moments of clarity, and for these we are grateful. But, since the images remain blurred, we should practice the Christian virtue of humility in what we claim to know and to have "right" in our historical perspectives.

This leads us to the most difficult question of all in deciding what difference it makes for a Christian to study history: Should Christians study God or humankind? Before we can answer that question directly, we must make a few preliminary points. As stated above, the book assumes the validity of Christianity as personal faith and as worldview; hence we seek an integration of that faith commitment with historical study. By historical study, we mean no special definition, unique to Christians, but that definition common to all people who study history. Christians study the same discipline as persons of any faith or of none. Because we believe in the coherence of truth, we want to have the broadest discussion of all reality with all persons interested in serious inquiry. Christians, therefore, should not try to redefine history.

In order to have an acceptable dialogue, all historians must discuss the same reality. Reality includes all past human activity. At a stroke, therefore, a bone of contention arises. Christians follow God, or, as some say, "belong" to God. Much "Christian history," i.e., the Bible, is a testimony to the acts of God. But, as historians, we study past human activity. Here is the contentious point: We historians study humans, not God. Historians with research degrees agree on this. I know of no working historian whose subject is God in history—I do not mean the idea of God in the idea of history, but God as known by Hebrews and Christians and history by everyone who does history. Occasionally a historian writes a providential history and Christian scholars find it unpersuasive. For example, historian John Warwick Montgomery wrote *Where Is History Going?* (1969), which sparked a fierce debate in the pages of *Fides et Historia,* the journal of the Conference on Faith and History, a professional association of Christian historians.

As historian Stanford Reid (1973) has suggested, we study humans rather than God because of the radical break between time and eternity. God, who is in eternity, is inexplicable in human terms. We simply cannot reason from our time-space to God's infinite space. We who can only partially comprehend what we call time can scarcely comprehend the One who clearly transcends time. Even Moses, who experienced a more direct contact with God than any person in recorded times, states flatly that "the hidden things belong to God." The twentieth-century Roman Catholic historian Christopher Dawson sees the Incarnation of Christ as the key event of history because it gives spiritual unity to the whole historic process. He states that providential events "have occurred as it were under the surface of history unnoticed by the historians." Thus, for historians to discern God's actions in modern history seems a sterile task because of the hidden nature of the subject.

At this point, caution must be sounded and a balance struck. Just because we can know little of God's intended purposes does not mean we can know nothing of them. Historian Frank Roberts summarized the twin difficulties of overassurance and overdiffidence: "The tendency toward overassurance has generally been marked both by its disposition to play down the complexity and ambiguity of history and by its inclination to emphasize the clarity of the divine plan and purpose in events of the past." On the other hand, overdiffidence inclines historians to reject a distinctively Christian approach to history as either impossible or undesirable. Some writers believe that, in the New Testament age, we can know nothing about the confrontation with the powers. Others assert that the historical method itself is incompatible with belief.

A balance between overassurance and overdiffidence is important. We accept the limitations noted above, not necessarily of a Christian approach to history, which this book will affirm, but of knowing the work of God in history and especially of "patterns" of providential action. The excellent book by British evangelical historian David Bebbington, *Patterns in History: A Christian View* (1979), develops a sensitive and penetrating analysis of the relationship between Christian commitment and historical study. Bebbington attempts to resolve the tension between "technical" history (the history that all historians practice) and "providential" history (the history that only Christians can know) by distinguishing between explicit and implicit renderings of faith commitments. Rather than a uniquely Christian history, Bebbington suggests that a believer can produce work that is "consistent" with a historian's Christian views. Christians can write a distinctively Christian product, but the Christian content will be implicit rather than explicit. I must say that I am not fully persuaded by Bebbington's conclusion that the reason for moving between technical and providential history is a tactical one, depending upon the audience to which the writing or the teaching is addressed. With academic colleagues one is implicit; with Christian sisters and brothers one is explicit. Yet, I appreciate Bebbington's work as the best statement yet on the calling of the Christian historian.

The Self-Consciousness of the Historian

By changing the focus from history to the historian, Christians can better understand their role as students of history and see more clearly their proper tasks. As the British biblical scholar Anthony C. Thiselton (1981) has suggested in a highly respected book on biblical hermeneutics, we must reconcile the relationship between the objective reality of study (the past) and the subjective beliefs we bring to the study of history. We have difficulty reconciling the two because our academic preparation does not encourage the attempt. Historian Oscar Handlin has clearly stated the problem in the context of studying European migration to America. In *The Uprooted* (1951), Handlin states that to understand the migrants he had to confront himself. Our academic preparation does not encourage us to confront ourselves, and, to be sure, that confrontation can be discomfiting.

What does it mean for a historian to be conscious of oneself before studying history? Perhaps an illustration will help. On Easter, 1977, the British Broadcasting Corporation televised a panel discussion on the subject of the Resurrection. Bamber Gascoyne asked a question of his fellow panel members that is of ultimate importance for Christian historians: If there had been photographic technology in place on the Emmaus Road, and if a picture had been taken of Jesus and his two walking companions, would that picture have shown Jesus of Nazareth, whom most people in Jerusalem knew? Or, did it require eyes of faith to see and recognize him in the breaking of bread? In short, if anyone could have recognized him, there is no need for an act of faith to know the risen Christ. If, as Christian tradition has it, we see him as the Christ by an act of faith, then we have to lay aside for a moment the objective reality of a person on the Emmaus Road and inquire into the subjective matter of how we develop eyes of faith. This change of focus from the thing observed to the observer often makes historians uncomfortable. Instead of discussing "reality out there" or "reality as it actually was," we are discussing ourselves—not the typical subject of discussion among us.

How do we develop eyes to see what we do see? More specifically, if we who are Christians wish to seek the application of our commitments in the actual doing of history, do we have eyes to see what others cannot or will not see? Historian Carl Becker, in his famous essay, "Everyman His Own Historian" (1935), makes the most cogent case for subjectivism. In Becker's view, the historian becomes the main focus of history. The past, he insists, is irretrievably lost; if it exists at all, it exists in the mind of the historian.

British historian E. H. Carr (1964) presents a more modest case for relativism and subjectivism. Carr sees history as a dialogue between the past and the present. "The past" is what happened, and that *is* lost, taken by itself. "History" is *our* reconstruction of the past. As in any dialogue, both parties bring their respective contributions.

Historical study, then, is subjective, to one degree or another. This came home very strongly to me in my first teaching job. We required a seminar of all senior students. At one session they read Bainton on Luther; the next, Erikson on Luther. The main instructor in the seminar invited me to participate in the latter session. He intended to portray Erikson as having "superceded" Bainton. I questioned this, and my colleague replied that Bainton merely gave a religious interpretation of Luther, whereas now "everyone knows" that religion is an illusion, not a "reality." He cited psychologist William James's *Varieties of Religious Experience* (1904) as evidence that one could not believe any longer in the normative and objective reality of religious experience. The students were uneasy about this discussion because we had ceased talking about Luther but had begun to talk about what my colleague and I knew about what everyone knew about reality.

Another example from a differing viewpoint may help us see the issue more clearly. Samuel Eliot Morison (1942) was criticized for one passage in his biography of Christopher Columbus in which he wrote that, on the first sight of land in the new world, Columbus "staggered" to the deck of his ship. The critic asked Morison how he knew that Columbus staggered. What were the sources? The ship's log records that on that day there were high seas and that the captain was ill. Very well, says the critic, but how do you know he "staggered"? Morison replied that he himself had sailed in a replica of the *Santa Maria*. In a high sea, a sick person in a ship like the *Santa Maria* does not walk to the deck, he staggers. In the end, Morison "knows" because his own experiences cause him to empathize with similar experiences in history. "Reality" then, one supposes, is what most of my friends and I know it to be. To those who affirm "I know that my redeemer lives," others will reply that they know that the class struggle exists. Is the historical task, in sum, an academic version of what John Lennon wrote for the Beatles: "I get by with a little help from my friends"?

Must we conclude then, that history rests on a thoroughgoing subjectivism, energized by self-authenticating experiences? Before we answer no, as we qualifiedly will, let us state clearly that, even if it were true, it would present no more difficulty for the Christian historian than for the Marxist historian or psycho-historian. Since the celebrated E. P. Thompson and Erik Erikson are allowed to see reality as they and their friends know it, why should Christians be embarrassed to see the past as we and our friends see it? What is sauce for the goose is sauce for the gander. While Christians do not necessarily endorse every disclosure of religious belief in history (some may not have been genuine), we nevertheless are open to an interpretation that affirms that, *in reality,* religiously motivated actions do exist. How do we know that Jesus came into people's lives and transformed them? We know because we, too, have met the Christ, whether as Catholics in the eucharist or as Protestants by making a "decision for Christ." The facts of history simply do not speak for themselves; historians speak for them from an interpretive framework of the ideas they already hold.

Objectivity Reasserted

Historian George M. Marsden (1984) attempted to keep communications between Christian and non-Christian historians open by appealing to the "common sense" school of thought, notably associated with Thomas Reid, the eighteenth-century Scottish philosopher. This helps us to account for the other side of the question, i.e., the experiences and understandings we have in common with non-Christians. Reid argued that the "first principles" that every human affirms do not depend on reason, commonly defined.

For instance, virtually everyone is forced to believe in the existence of the external world, in the continuity of one's self from one day to the next, in the connection between past and present, in the existence of other persons, in the connections between causes and effects. . . . In practice, normal humans simply find it almost impossible not to rely on basic means of gaining access to knowledge. Only philosophers and crackpots, he was fond of saying, would seriously argue against the reliability of these first principles. And even skeptical philosophers duck when they go through low doorways. So do Hindu mystics.

This common sense approach (in its technical meaning) makes very good sense, in the ordinary meaning of that term. Reid simply disposes of the philosophical concept of "ideas" and starts with common sense, which tells us that we can know directly something of reality. Knowledge, then, is not confined just to ideas but involves what is really "out there." There is no theory-independent access to events, no knowledge that cuts across all theories. So, in fact, there is some common ground of inquiry into the human past. In theological terms, this is referred to as *common grace.* Has not God created a coherent universe? Given the good will of historians, can we not communicate fairly well with the assurance that we are talking about the same things?

So where does this leave writing history from a perspective, Christian or otherwise? Again, George M. Marsden helps us with an analogy from gestalt psychology. Surely everyone, at one time or another, has seen the picture in figure 1 (see page 10).

At first glance most people see the old lady, and only later do they see the young lady. In the common sense understanding of things, both ladies are there, but not everyone can see them. The presenter of such a picture will not get his viewers to see the second image through argumentation. Seeing, and believing, that the young lady is there will come sometime as insight and it will change the viewer's understanding of what reality actually is. Christian seeing and believing is something like that. It is not that we see everything differently from non-Christians. All humans, as Reid pointed out, know the signs of everyday life. We come to

Figure 1. Young lady or old?

know God and God's work in a moment of shattering insight, flowing from things we have seen before many times. We know and experience grace in worship, in nature, or reading a Scripture passage. When that understanding comes, we say, "Oh, now I see," and a pattern of reality emerges from what was there but unseen.

Historians should have little difficulty understanding this because it is like the way we actually do history (minus, of course, the soul-shattering nature of the insights). A metaphor, again, will help to clarify: Access to reality is limited by a series of lenses like the multiple-lens glasses that eye doctors test us with. While it is true that each person wears a different set of lenses, most normal people can read most of the letters on the chart. As Christians we have an extra set of lenses, which perhaps allows us to see what others see but also more than they and perhaps more clearly. As American philosopher Nicholas Wolterstorff (1976) has argued, these extra lenses can act as "controls" on what we see through ordinary lenses, insuring that common sense beliefs will not contradict special beliefs.

In conclusion, I hope the balance is now clear. On the one hand we do history from a perspective, and the adequacy of a historical interpretation must take into account the historian who is in dialogue with the past. Nevertheless, all reality is not mere opinion and private experience. As Christians we say we see not the antithesis of what non-Christians see but all that they see, and more, because we have that extra set of lenses. Further, that extra set of lenses not only helps us to see more, it helps to order and to control our understanding of what comes to us through the ordinary lenses.

A Whole View of Reality

Canadian historian C. T. McIntire (1981) has helped by providing an explanation of this relationship. Everyone knows that reality consists of time and space. Christians and non-Christians alike see these two dimensions. We Christians insist that there is a third dimension—spirit—and that a whole view of the world must be three-dimensional. Moreover, these dimensions are not arranged in a hierarchy of importance; rather they are integral to each other. Now, secular-minded people may well object to the claim of a spiritual dimension, but even they must see that good and evil exist in the world and that a spiritual dimension refers to something more ultimate (even if ultimacy is not God but possibly even merely the mode of production). McIntire uses more comprehensive terms for time, space, and spirit—historical, structural (ontic), and ultimate.

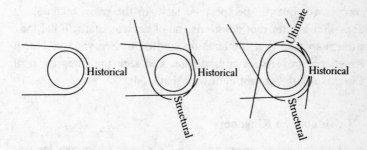

Figure 2.

Presented in this light, the integral and fully three-dimensional world on which Christians insist, is something that we Christians see at first, but it is not so far from the experience of non-Christians that they cannot understand it. Perhaps like our dual gestalt picture, if it is pointed out that the nose of one is the chin of the other, people will see what we mean. When we insist that a Christian worldview is a whole view of the world, non-Christians need not dispense with everything they know but realize that the extra set of lenses alters our perception of reality.

Of course, it needs to be said in closing that Christians must act "Christianly" toward others in discussing these matters. We who say we supposedly have the best view of reality must not come to others in triumphalism. Modesty and humility are becoming traits. Even if we have all the lenses of our glasses on fully, we still see through those glasses darkly. And when we become arrogant and militant we must recall that Jehovah sits in the heavens and laughs when we imagine vain things.

Finally, as author of this book, let me say a few words on how I expect it might be used. The premise is that students in "Western Civ" courses will read a "standard" text provided by the various (presumably "secular") publishers. This book is meant to supplement that reading, not supplant it. We intend to analyze from a Christian perspective certain issues the text may have discussed factually. This book does not intend to stand on the sidelines and complain about the perspectival lacks in the main textbook. It does offer a pattern of questions and of analysis that will help the student to take that information studied by everyone in North American universities and colleges and seek the integration of Christian faith commitment with that study.

Crisis and the Kingdom

This book has an orientation, and it is best to acknowledge it at the beginning. In every time and place, and in every person, there is a struggle between "the real" and "the ideal." Every person and every society has a philosophy of life, a *worldview,* set in ideal terms, of what "ought" to be. Also, every person and society experiences life-as-lived, which is always somewhat different—sometimes markedly different—from life-as-hoped-for. The "is" always falls short of the "ought." This is part and parcel of the human condition, acknowledged by writers everywhere, echoed by St. Paul, who spoke for Christians of all ages in confessing that while he knew the ideal requirements he frequently fell short.

So, this is taken as self-evident for the human condition. Moreover, humans can, and do, get along in their personal and social lives, somehow accommodating to this gap between realistic and idealistic formulations. Yet, a crisis of behavior and belief ensues when the gap—always present—grows too great, when the realities of life-as-lived grow too distant from ideal formulations of what life "ought" to be like. It is always events that require ideal formulations to be questioned, then reformulated. It is always the crushing weight of events that demands that people admit that they cannot go on with their previous view of the world. It is always in times of crisis—not in prosperity—that ideologies are tested. As those with even the merest acquaintance with personal grief will attest, it is not on a lovely summer day by the seashore or in the mountains that we come to grips with the meanings of life but rather standing by a loved one's hospital bed or burial place. What we know personally is also true societally. It is not in times of peace and prosperity that our ideologies are tested but during times of war, massive unemployment, famine, or plague.

The idea of "crisis," then, will be a guiding principle of the argument in this book. It is not only helpful because it describes the human experience but also because it releases us from the assignment of giving praise or blame. Historical study is about human understanding, and we do well to avoid a judgmental attitude that seeks to praise or blame. Empathy, rather, is a human (and preeminently Christian) attitude that one does well to employ in all aspects of life, especially in historical study. This book invites the reader to apply to historical study what he or she already knows (and what St. Augustine said so long ago), that the kingdom of God and the kingdom of the world are not the same. While we may dwell in both for a time, we know which one is coming and which one we should seek first. It is with this "kingdom vision" that this book is written, and, the author hopes, it is in that vision that it will be read.

Chapter 2

GREEKS AND HEBREWS:
TYPOLOGIES FOR WESTERN
CIVILIZATION

Much of the tradition of thought and experience that Western people hold dear began in Greece, specifically Athens. No serious student would quarrel with this assertion. However, less consensus exists as to the foundational role of the Hebrews. They were, after all, only a small tribe among many in the Middle East, practicing a peculiar monotheistic religion. Indeed, one might not be particularly interested in them for themselves had not their religion been the basis for what became the dominant religion of the West—Christianity. However, because Christianity is also formative in the development of the Western heritage, what we said for the Greeks we also say for the Hebrews, that much of what Western people hold dear began with the Hebrews. Further, this book—premised, as it is, on a Christian perspective—sees the Hebrews not just as the providers of one of the foundations of Western society. It sees the Judeo-Christian tradition as providing principles that might guide our understanding of the main story of development from the Greeks.

The main textbooks available to students of Western civilization do a competent job in describing the outline of developments in Greek civilization. Consequently there is no need to more than repeat here that the civilization began with the Minoan and Mycenaean cultures and was extended to the Greek mainland and throughout Asia Minor in the "Homeric" period and that the founders named their country Hellas, after the mythical forebear, Hellen. The development of the city-state is one of the great Greek achievements. And while Athens and Sparta may be best known, there were other important ones—Ephesus and Corinth—as readers of the epistles of Paul will recall. Yet, it is Athens we rightly remember most readily, for in Athens more creations of the human mind occurred than in any other place in the history of the Western world.

The Hebrews, for their part, are less well covered in standard textbooks precisely because of their obscurity. If one sees the early story of Western civilization, as most textbooks do, as proceeding from Mesopotamia, through Egypt and Persia, through the Greeks and Romans to Europe, then the Hebrews *are* a relatively unimportant part of the main story. We see in the experience of the Hebrews an interesting alternative example; because of the Hebrews' impact on Christians, we are interested in that example. In doing so, let us be aware that we *are* stepping outside of the main story line of Western civilization, as some textbook authors see it. For us to stress the importance of the Hebrews as an alternative to the Greeks is to offer an alternative to the main convention of telling the Western story, i.e., that history is the development of democratic liberty and individualism. For us to see value in the example of a people who defined themselves as faithful to covenant and community is to suggest that there was—and is—a framework for society other than "secular humanism."

Over the next few pages I would like to lay out two typologies for society. Let us be clear that a *typology* is not a picture of a society but an analytical construct of our own in which we organize certain themes of a society for purposes of our own understanding. While it is helpful in disclosing truth about the society, it may not be a fully accurate picture of that society. Nevertheless, typologies are helpful if we remember what they can and cannot do. I would like to use two words, selected for their functional worth, that do not prejudice our interpretation from the beginning. The words are *tribe* and *polis*, which relate, respectively, to the Hebrew experience and the Greek (i.e., Hellenic) experience. Again, while not a perfect rendering of reality, they allow us to begin to accomplish the purpose of this book—to think self-consciously as Christians about history.

Greek civilization was both of long duration and rich texture. So when we say *Greece* or *Athens* we must say what we mean. Remembering our format of ideality and reality, the ideal-types of Greek civilization came to expression in Hellenism. To understand it we turn to one of the foremost historians, Arnold Toynbee (1959), to give us a summary word-picture of Hellenic life. It was centered on the city-state (in Greek, *polis*). The polis, however, had existed before Greek ascendency, most notably among the peoples of the Tigris and Euphrates rivers. So, while the polis is not unique to Greece, it is one of the main features of Hellenic civilization. What *is* unique about the Greek polis is that it was a means by which practical expression could be given to a view of life. The Hellenic philosopher Protagoras first wrote a phrase to which we shall have to return many times in this book: "Man is the measure of all things." As Toynbee writes, if we were to take that phrase out of Greek language and put it into Judeo-Christian language, "we should say the Hellenes saw in Man 'the Lord of Creation,' and worshipped him as an idol in the place of God." We might recall that this form of "idolatry" is not unique to Greek civilization, because, in one form or another, it has been one of the dominant philosophies throughout history in Western culture. For present purposes we should note that Greek culture is significant in its devotion to this worldview. Because of the continuing impact of Greek thought in the West, it is of great importance to understand its origin.

But before developing the Greek type of society fully, let us turn briefly to the Hebrews. While not discussed as much in textbooks, the main contours of Hebrew culture should be fairly well known to readers of this book. The Hebrews were—and are—first and foremost a people of a story. What held the Hebrews together was not so much a sociocultural style or geographical place, because those varied; rather, it was an affirmation about God, about humankind, and about the nature of the universe. The Hebrews were radical monotheists who founded their personal and collective identities on an assertion of belief that their God was the creator and sustainer of the universe. Their God, Yahweh, existed before what humankind calls "time," and, out of nothing at all Yahweh created everything that was. The most essential point of their worldview, therefore, is that everything begins and ends with God and that all humans should take an appropriate (i.e., deferential and obedient) place in relation to God. To do any other is idolatry. First in oral tradition, then later in sacred writings, the Hebrews maintained that story throughout their existence, whether settled in one place or wandering in several places. Some of the later Judeo-Christian sacred writings, indeed, took note of the difference between Hebrew and Greek ways of belief and behavior. For people to believe and behave as the Hebrews did, believing in a God who not only created them but loved them and intervened for them, was, in Greek eyes, foolishness. As even a quick reading of the letters of Paul will attest, the Hebrews and the early Christians were aware of the considerable differences between Greek and Hebrew foundations of life.

While it might be stretching the point a bit, it is possible to organize Hebrew and Greek thought around two poles of consciousness—what a twentieth-century philosopher called their "ultimate concern"—Yahweh glorified and "Man" glorified. While this chapter will end by wondering if these distinctions are too neat and tidy, they help us to see what real differences there are in foundational worldviews and how those differences have continued throughout Western history. In the following few pages we will ask five questions and then discuss the differences between Hebrew and Greek responses, remembering that the functional terms *tribe* and *polis* can be useful in developing the typologies we discussed earlier.

How Do People Belong?

The Hebrews are most tribal in their sense of group membership. As in any tribe, one cannot *become* a member; one *is* a member. The Hebrew story is one of exclusive belonging in that group, which is held together by a primal sense of kinship, with much emphasis given to a covenant based on blood relationships. Moreover, this people is a peculiar delight to, and concern of, Yahweh. Since he protects and guides them, sometimes as residents on a land that Yahweh gave to "Father Abraham," sometimes wandering, or sometimes in captivity, they never wavered in their sense of peoplehood. Indeed, the common story they told their children in future generations helped bind them together all the more—the affirmation that God had been both refuge and strength from one generation to another. Later on, of course, after the life and death of Jesus of Nazareth, a considerable disagreement ensued among Hebrew Christians as to whether or not non-Jews could also be heirs to the covenant promise given to Abraham. We shall return to that later on. But for now, it is enough to say that even after Peter's vision and the bringing of the Judeo-Christian gospel to the Gentiles, there was—and is—an exclusive covenant bond between God and his people, sealed in blood.

The Greek way of the polis was, at first, superficially similar to the Hebrew way of tribe. Athenians could supposedly trace a lineage to common ancestors, and they were tied into a community by a common and ritual deference given Athena, the patron-goddess of the city-state. They also believed in community in that they did not regard the government and the services of the city as remote from themselves. But, importantly, they developed a sense of citizenship that would have been foreign to the Hebrews. Citizens had certain rights, guaranteed by law. Moreover, while citizenship was naturally conferred on those born in Athens, one might move to Athens from elsewhere and, in due course, become a citizen. And if one moved out of Athens to a colony, as many people did, the "mother" city of Athens might have kept political control but it did not extend its citizenship. Rather, the emigrants became citizens elsewhere. While citizenship was important in community-building in Athens, it did not function with the depth and duration that membership in the covenant people did for the Hebrews.

What Are Their Economic Activities?

The Hebrews were people of the land. To be sure, Jerusalem was a considerable city, but the main experience of the Hebrews was on the land. Like most tribal peoples, they were able to maintain their religious and ethnic exclusiveness by keeping their distance from the "secularizing" and "modernizing" tendencies of the city. They were principally engaged in tilling the soil and tending their flocks and herds. People in towns did make and trade goods, but the countryside was the center of Hebrew tribal life. Indeed, one of their main prophets, Jeremiah, deplored the growth of city life, believing and asserting that "death" was in the city, while "life" was on the land.

The Athenians were at first similar to the Jews in that they, too, were agriculturalists. Right down to the time of Pericles, a majority of Athenians owned some land. But, while agriculture never died out, the rise of Athens as a commercial center shaped its destiny more than agriculture. In fact, by 350 B.C. Athens imported about two-thirds of the grain it consumed. The development of commerce, and the cash economy related to it, helps explain the ascendancy of Athens. Cities—especially fairly large ones—by their nature cannot be agricultural centers but commercial centers. And as cities grew, like Athens did, citizens moved there in pursuit of commercial activities.

What Is the Place of Religious Values?

For the Hebrews, religion was not a part of life, it *was* life. The Hebrew Scriptures resound with that point, that God and his commands cannot be escaped even if one went to the farthest part of the sea. Yahweh was the beginning and the end; even though the realities of Hebrew life were sometimes at variance with this assertion, the ideal formulation of life always had God as central. It is impossible to think of the Hebrews in any other way because, without their religion, they would have had no identity and because their religion sustained them throughout their many difficulties. Religion was not simply a common value for the Hebrews; it was a core value.

For the Greeks, religion was also very important at the beginning. They had an origin myth and a belief in gods who intervened in the affairs of humankind. On the top of Mount Olympus, the highest point in Hellas, dwelt Zeus, who presided over a family of lesser gods. Zeus (Jupiter to the Romans) had special oversight for the sky. He delegated special responsibilities to his brothers: Poseidon (Roman: Neptune) ruled the sea, and Hades (Roman: Dis) ruled the underworld. The Greeks perceived these gods in anthropomorphic terms; they were very "human" gods in their closeness to mortals, unlike the "wholly other" God of the Hebrews. Indeed, the favorite gods in Athens were those with whom citizens could readily identify: Aphrodite (Roman: Venus), the goddess of love and beauty; Ares (Roman: Mars), the god of war; Hermes (Roman: Mercury), messenger of gods and patron of athletes; and, most of all, Dionysos (Roman: Bacchus), the popular and beloved one because he was the offspring of Zeus himself and a mortal woman, Semele, and thus even more accessible to humankind. He became associated with the twin ideas of wine and indulgence (although more so with the Roman appropriation of Greek religion, hence the term *bacchanalia,* or orgy) and of fertility, especially in nature, of birth, maturity, death, and rebirth. But it is important to note that Greeks had no dogma, or official beliefs, no Scriptures, and no official class of priests, thus making religion fairly democratic.

All the above notwithstanding, the Greeks are most significant in the way their intellectual leaders increasingly rejected religious attitudes. Thus, as one major textbook in Western civilization puts it, "They made the intellectual leap from a primitive view of nature to a reasoned, analytic view" (Greer, 1987). First in the Sophists, especially Protagoras, and later in the plays of Aristophanes, the poems of Euripides and, above all, the philosophies of Socrates, Plato, and Aristotle, the "Greek mind" developed that "demythologized" the world. While not necessarily attacking God (or the gods), though Aristophanes used ridicule aplenty, the result was the elevation of humankind in the pursuit of what they called the excellent, or virtuous, life.

What Is the Role of the Individual?

The Hebrews had little appreciation for individualism. They understood, as Christians since then have understood, that God comes to persons, not groups, at least in the first instance. But—and this is a crucial point—God does not intend that persons remain as individuals. They were once individuals, but their new identity in God requires that they become part of the people of God. In short, in the Judeo-Christian scheme of things, individuals do not pursue that which is right in their own eyes. Rather, individuals are to become one people in Yahweh, and the sacred writings abound with images of intimate union (vine-branches, head-body) in which all persons must take into account each and each all.

In social formulation, then, we see a conflation of selfhood and society. The "healthy" self for the Hebrews was a person whose concept of selfhood did not insist on a distinction between self and society but asserted that self and society must merge. It was seen as socially deviant (because religiously deviant, given the central position of God-consciousness among them) for a person to insist on his or her own way or even to suggest that there were "rights" for the individual to safeguard a person from the encroachment of the community. The values of the group over time ("from one generation to another") are constant in the devotion to God and the determination to follow him.

The Greeks were markedly different in this respect. The life of the polis in Hellenic culture was founded upon a belief in the good life of rational, moderate, and virtuous behavior and belief. Moreover, this good life was not for the few but the many (though not all; there were slaves). These citizens could only enjoy this good life if the community and its institutions gave the individual some considerable personal freedom. Their highest ideal was the application of reason—although in moderate, "golden mean" terms—to all things.

In societal formulation, then, we see a differentiation of self-hood and society. The healthy self for the Greeks was a person who insisted on a distinction between self and society. It was socially deviant to desire to merge one's identity with that of the group because to do so would have been self-denial, even self-annihilation. Since the only life worth living was the one in which a free individual examined it for himself or herself, the ideal formulation of Greek society assumed the primacy of the individual. Especially since there was no God (or gods) who demanded unconditional obedience, the Greeks could encourage people to found their lives on this relatively optimistic view of life.

What Is the Political Theory?

The Hebrews gave little attention to their forms of political organizations: Sometimes there were kings, while at other times there were charismatic leaders who ruled. But whatever the forms of political organization, there was a constant theory that, given what we have already said about the Hebrews, should be quite obvious. Their society was one in which authority was clearly understood. Authority resided in God alone, and with this undivided sovereignty, there could be no suggestion of people circumscribing or limiting that authority. When those in authority said, "Thus saith the Lord," those dicta were not up for majority vote. When Moses came down from the mountain, "the Law" was not ratified by the people or by legislatures; when Job suffered greatly, his part was not to question but to obey. To be sure, the prophets frequently called upon the ruling authorities to act justly, especially to the poor. But that was not out of a democratic attitude that cried out "power to the people" but rather out of a foundational conviction that God has a purpose for humankind and that that purpose should not—indeed, could not—be opposed by secular rulers. Authority rested, in the end, in God alone and in those he anointed to rule.

The Greeks saw things differently. Athens was far from a perfect democracy, in that only free males were citizens. It nevertheless developed over time a system of rights for individuals that was organized by "law," which was not merely "given" but to which citizens had access and over which they had some control. Whether in the aristocratic organization under Draco or the democratic organization under Pericles and Cleon, the Greeks insisted on a kind of democratic ideal that would have been foreign to the Hebrews. Whether in *The Republic* of Plato or the *Politics* of Aristotle, they tried to reconcile the perennial problem of "the one and the many" by insisting that if the right people (those with "virtue") ruled well (never seeking their own regional or class interests but that of the commonweal), society would be well served, and legitimate interests of the group and of the individual could be reconciled in the good life here on earth.

Having contrasted Hebrew and Greek ways in terms of several important questions, it is now possible to make some conclusions. The Hebrews were "tribal" in that they were organized according to kinship; they engaged in primarily agricultural pursuits; religious values were core values; the group had primacy over the individual; authority was undivided. The Greeks represent the type of the polis in that they were organized in terms of citizenship; they became involved in commercial activities; they became functionally secularized; they celebrated the individual; they divided the sovereign authority of the state and guaranteed rights by law. Figure 3, while far too simple, helps us to punctuate this typological difference:

HEBREWS	GREEKS
Tribe	*Polis*
Kinship	Citizenship
Agriculture	Commerce
Religious	Secular
Community	Individual
Authoritarian	Democratic

Figure 3.

While this representation of types does some violence to reality because it makes life too neat and tidy, it helps us organize some of our ideas about the twin foundations of Western civilization and recognize that those foundations are, not only different, but are at considerable odds with each other. Moreover, it helps us to see graphically that these various points of distinction between Greeks and Hebrews do not exist in isolation but exist in affinity. It is possible for democratic liberty to exist in a society that celebrates the individual citizen living in the secular city. It might only be possible for authority to be undivided in a sovereign God and his appointees in a society that elevates the community of kinship-related people whose identities are founded on core religious values, living in obedience on the land that the Lord God gave them.

But we must also be careful to note what we cannot learn from these typologies. Some Christian writers in our time would make much of what we have said and then invite people to choose this day whom they will serve, God or humanity. They frequently use a term—the *antithesis*—to describe the difference between divine and human thinking. Their plausible argument runs roughly like this: God's ways are not humanity's ways, as witness the Hebrews and the Greeks. And, since Western society is in a dreadful predicament, we must see that evil befalls a society that "abandons God" at the center and "elevates Man" to the center of a worldview. Both as personal faith and as worldview, "secular humanism" in all its manifestations must be rejected by Christians who have an "antithetically" different foundation for life. This argument carries special weight in North America because many Christian colleges found their reason-for-being on the assertion that their curriculum is "Christ centered" and, presumably, essentially different than that of "secular humanist" universities and colleges. Indeed, many of the readers of this book will find themselves precisely in this situation.

While the above line of argument is plausible, it is not persuasive. To be sure, in the mind of God and at the end of time, there is and will be what C. S. Lewis called "the shocking alternative of a confrontation between the two kingdoms." Any persons who would think "Christianly" must affirm that. Yet—and this is vital to the remainder of this book—the historical task is not in the realm of defending faith. Historians—professors and students alike—seek to understand human activity over time. While, as suggested in the previous chapter, our presuppositions do inform the pattern of our understanding, they do not prejudge the story.

Let us return for a moment to the graphic that delineates in ten words the differences between Hebrew and Greek "types." Are Christians in North America ready to abandon democratic and individual liberty in favor of a communal authoritarianism? Are Christians in North America prepared to apply to their nation-states what was meant in the Bible for Israel? If we who are Christians are attracted by certain Hebraic traits (religious values, community support, the undivided sovereignty of God) *and* some Hellenic traits (citizenship, individualism, democracy), how can one say that we should live by "the antithesis"?

It seems that it is precisely *the wrong question* to ask of history what we cannot do in our own lives, regrettable as that might be. Remember our previous discussion that "the past" is what happened, whereas "history" is our engagement with the past. When we engage the past in Western civilization we do, in fact, see people trying to reconcile (to "synthesize") what apparently cannot be reconciled. Yet they—and we —try to do what, apparently, cannot be done. Historical study must rise above assigning praise and blame, either to historical figures or to ourselves. The story of humanity in the West is a story of trying to bring together what St. Augustine called the two cities, of God and of "Man." Even though we might *know*, propositionally, that this is an impossible task, it *is* the story we observe and in which we ourselves are involved. It will do no good to shout at historical characters—or at ourselves—"Don't do it," because they—and we—live in "the already" and "the not-yet." The kingdom for which we pray *has* come but still it is also *yet to come.* Whatever normative satisfaction we may draw from learning about a "radical antithesis," real life is lived in the middle ground. In this book we will see humankind struggling between the dual imperatives foundational in its culture. At certain times one will be ascendant, at other times the other. The end of historical study, after all, is self-understanding. As we develop empathy for those who struggled within the Western duality, we may see ourselves, and our own time, more clearly.

Chapter 3

THE HISTORICITY OF JESUS:
A PROBLEM OF ''TEXT'' AND ''MYTH''

Once I was asked if, as a historian, I thought the Genesis story to be "history like anything else is history." I replied in the affirmative, but then I asked my inquirer to refine her own question and to tell me what it means to say "like anything else is history." She said, "Well, I suppose, that it is for real." What does it mean to say that the Garden of Eden or the Bastille are "for real"? And, important to this chapter, what does it mean for us to say that Jesus Christ is "for real"? Apart from the fact that the church over time and Christian leaders in our time have replied to the question Jesus himself asked ("Who do men say that I am?") by repeating Peter's testimony ("You are the Christ"), what can we, as students of Western civilization, say about Jesus of Nazareth? By now, readers of this book will have become familiar with the definition of history being employed here: The "past" is what happened; "history" is our engagement with the past. So, in order to determine the historicity of anything or anyone, it cannot be history if *we* do not engage it. While we might use scientific methods of inquiry (we don't accept just anything from the past), and while we are fair and judicious with our use of sources (for honesty's sake we wouldn't "cook" the evidence), we need to confront ourselves and engage our subject or we simply are not "doing history."

History and Faith

The Christian faith is precisely that, a "faith," by which we mean that we offer into evidence things that are not seen in the "normal" way of seeing (recall the introductory chapter about Christian ways of seeing and knowing). For those of us with eyes of faith, the answer to the question about the historicity of Jesus is rather straightforward. With the Church of all times and places we reply, in the words of the creed that comes from the apostolic age, "I believe in God . . . and in Jesus Christ, his only Son, our Lord." To persons without eyes of faith, these words will mean little, and they might note that we are not saying our own words but the formulaic words of a believing community. Our own words come a bit more self-consciously, as we grope for the right formulation of what we believe. All Christians believe the same thing, that "Jesus Christ is Lord." But they express it differently: Those raised in a liturgical church expression might say, "I have met the Christ in the eucharist," while those in a Free Church expression might say, "I have made a decision for Christ and now he lives in my heart." But questions persist, and these are not badgering but honest questions: How can someone know that Jesus is Lord without an act of faith? Or, is there any way, beyond the believing community's testimony (whether in the year 300, 1500, 1800, or 1990) that they know by faith? What are the sources?

The factual, "historical" reality of Jesus of Nazareth is beyond doubt, except one matter to which we shall return. Most people agree that Jesus was born in or near Bethlehem in, more or less, the year Christians have historically assigned when they renumbered the calendar. It is also clear to most people of any faith, or of none, that he was raised in Nazareth and that he lived a quiet and unexceptional life with his parents and siblings, this despite the great ferment going on in Roman Palestine at the time. We know that he was a very religious Jew and that he was circumcised in the normal way, among other ritual duties that he performed. Sometime around his thirtieth year he began to have a public role. He, like many other charismatic preachers in his time, gathered some followers and moved freely around Palestine, quoting the Hebrew Scriptures and preaching that (what he called) "the kingdom of God" was at hand. Indeed, in that respect, he was not markedly different from some other itinerant Jewish preachers of his time. In a moment of heightened sociopolitical disturbance, he was seen—by Roman and Jewish establishments alike—to be disturbing the peace. Again, like many other charismatic teachers, he was arrested and later put to death by the relatively common method of crucifixion. This story of Jesus' life is widely agreed upon, and, with minor exceptions, no one seriously dissents from so unexceptional a story. On this level, the historicity of Jesus is undoubted, and, in the words of my questioner, noted above, Jesus is "for real." But if this were all there was to the story, no one would care very much, and there would be no Christian religion.

The distinguishing mark of Christian belief, of course, is not the prosaic story outlined above but the assertion that this same Jesus, who died in the way of all flesh, did not remain dead. He is said to have come back from the dead, to have "risen," and later (having spoken with people who were "for real") he "ascended" to heaven where he now exercises lordship—while somehow, at the same time, living in the hearts of those who know he lives. Now, this *is* extraordinary. The "historical Jesus" surely exists, i.e., the Jesus of Nazareth who lived and died in Palestine during the Roman era. But, a person rising from the dead in historic time and living ahistorically both in "glory" and in peoples' hearts, that *is* controversial. With this we come to the difficult part of assessing whether or not Jesus is "for real," if by that we mean to ask, "Is there any way independent of the testimony of the believing community that we can answer the question?" By "the testimony of the believing community" we mean to include its main source, the Bible, as well as the unbroken train of testimonies of believers, from the earliest times of the Church to the present day. The answer, truth to tell, is *no,* there is no independent body of information that will corroborate the story as told by those who have a personal stake in its outcome. In short, there is no way to satisfy the person who, like ("doubting") Thomas, would like to believe but who cannot believe without irrefutable, empirical evidence.

At this point in the conversation, Christian believers often shift the discussion to their source, the Bible. You *can* trust the Bible, they say. Some Protestants even ground their religious identity in such terms, calling themselves "Bible-believing Christians." Catholics would not typically take such a stance. But the Bible, while surely trustworthy, is not without its difficulties, and it does no good for Christians to deny these difficulties. The principle problem is that all of the New Testament was written after the fact. However one dates the Gospels and Epistles, and however one sorts out the matter of authorship (and there *are* many problems in all of that), it remains clear that the New Testament writings themselves cannot be regarded as independent testimony because they were written some years after Jesus' death, with the Gospel of John (the most "messianic") the last to be written. Moreover, they were written by people with an "Easter faith," that is, with a coherence building upon an interest in the story agreeing with a developing faith. One of the most arresting examples is the prayer Jesus himself taught, commonly called "the Lord's Prayer." In all reliable, early manuscripts the prayer ends with "deliver us from the evil one," but later versions of the manuscripts added "for thine is the kingdom, the power, and the glory."

This point is not meant to cast doubt on the Bible, but it is meant to suggest that an appeal to the Bible does not, in itself, resolve the question about the historicity of Jesus as the risen Lord. The plain fact remains that the main evidences we have for Jesus' messiahship and for the Resurrection are not simple chronicles or accounts but interpretations of certain events by writers who may or may not have been witnesses to the actual events and who, in any case, had an interest in perpetuating belief in the claims surrounding those events. The "quest for the historical Jesus" is a real quest that has animated the work of generations of scholars. What we can say, with intellectual honesty, is that by normal canons of historical methodology Jesus of Nazareth is historical. But Jesus Christ, "the risen Lord," cannot be documented as "historical," normally defined. This is not at all to assert that Jesus as Christ is not "for real." It is to say that we join the testimony of the believing community in affirming the central tenet of our faith; but that is exactly the point, it is a *belief*, founded in a *faith*, not a conclusion induced from indisputable "facts."

Engaging the Person

Having said that, moreover, we must persist in asking again just what history is. To repeat: History is *our* engagement with the past, and we engage the past from where we are. Many historians—including the present writer—reject the claim that reality can only be known through natural, empirical reason. Whatever the subject in question may be, we insist on ways of knowing other than the Greek ideal of applying reason to all things. If we are Christians, we have already accepted ways that truth comes to us other than through reason. There is an existential sense of knowing, for example, what we describe in our own special language as "the ministry of the Holy Spirit." We will not accept the rejoinder that such things are too purely "subjective" to be trusted. We say that all stories—in the act of remembering—are subjective. We assert as normative that, as was mentioned in the introductory chapter, the proper mode of historical study for Christians is the integration of time, space, and spirit. So it should not be a cause of embarrassment to us that we say "Jesus Christ is Lord."

Now, a careful qualifier must be inserted. If I asserted the above statement, and only I, then the charge of subjectivity would have great weight. But I—and we—find ourselves in a long history of such a testimony in the believing community. So our affirmation is not something of our own invention peculiar to our time or our own personal histories. But, a critic persists, "You are just repeating a myth." Yes, we are, but we need to define what a "myth" is. There is a sense, of course, in which one can dismiss a myth by saying it is untrue. But in the fuller sense of myth we mean a story. It is a story rooted in the past, but it takes an aspect of the past, saying that aspect is "the key" to understanding the past. Moreover, it is not purely private but a story shared by a large number of people, and—importantly—over a long time. When a community is built around a common story, and when those community members gather periodically to celebrate that story, and when that story helps them to explain the past and invite the future—then people are believing and acting "mythically." In this technical sense of the word *myth* (not the thing untrue but the thing believed in by a historical community), the Christian faith—centered on the "risen Christ"—is "mythic." Latter-day inquirers, like ourselves, engage the story for ourselves in various ways: through the acts of Christian worship (singing, preaching, and the eucharist); through the testimony of the whole Church (the creeds); through our own experiences (conversion).

By acting and believing in this way, we are "doing history," engaging the past. And by engaging the risen Christ, we are doing something like what we do in historical studies of any kind. This particular case is unique because the person we engage in the community of belief is, we say, no less than the Lord of life. For us that makes all the difference. There is no better formulation of the tendency of my argument here than that of C. S. Lewis. So, rather than paraphrasing Lewis, let us conclude this chapter on the problem of the historicity of Jesus by quoting Lewis in a familiar passage from *Mere Christianity:*

I am trying here to prevent anyone saying the really foolish thing that people often say about Him: "I'm ready to accept Jesus as a great moral teacher, but I don't accept His claim to be God." That is the one thing we must not say. A man who was merely a man and said the sort of things Jesus said would not be a great moral teacher. He would either be a lunatic—on a level with the man who says he is a poached egg—or else he would be the Devil of Hell. You must make your choice. Either this man was, and is, the Son of God: or else a madman or something worse. You can shut Him up for a fool; you can spit at Him and kill Him as a demon; or you can fall at His feet and call Him Lord and God. But let us not come with any patronizing nonsense about His being a great human teacher. He has not left that open to us. He did not intend to.

Chapter 4

THE EMPIRE AND THE CHURCH: THE PARADOX OF THE TWO KINGDOMS

The Romans organized the Western world. We cannot really think of the whole idea of a Western civilization without the achievement of the Romans. There were many peoples and tribes in what we now call the Middle East, North Africa, and Europe, but it was the Romans who brought some sort of unity to the totality of "the West." Paradoxically, however, it was within the Roman Empire—surely a "kingdom of this world"—that the new religion of Christianity developed, grew under persecution, then flourished as the religion of the empire. Under the Roman auspices, along the Roman roads, using the Greek and Roman languages, Christianity went forward and became a culturally dominant way of thinking in the West. In this chapter, we want to think about the Roman achievement and the Christian success.

The Roman Conquest

Rome was a small entity centered on what we now call central Italy. Its initial sphere of influence was "Latium," a coastal region of about 150 miles, roughly from a bit north of the city itself down to Pompeii and the area of Mount Vesuvius. As Romans gradually moved into the southern part of the peninsula, they encountered Greek culture in "Greater Greece" (*Magna Graecia*). Sensing a more elaborated ("higher") culture, they began to absorb many aspects of Greek culture into their own, from the alphabet to the gods. Most interestingly, the Romans adopted Greek sociopolitical organization and thinking. They established a "republic" (*res publica,* or "commonwealth") in which the Greek ideals of law, balance, and moderation were attempted.

Military victory followed victory for the Romans, although some came at great social and economic cost (especially the two Punic wars against Rome's Mediterranean rival, Carthage, in the late third century B.C.). Later, having conquered the Greek empire in the east, the Romans were largely in control of the Mediterranean and its reaches. Yet all of this conquest had gone on without a general plan. Drained at home through costly victories and stretched abroad by having to rule conquered provinces, the Romans experienced great stress on their republican institutions in the last decades of the second century B.C.

The demise of the Roman republic—or better, its transition into the empire—is explained both by social stresses at home, especially the distinction between patricians and plebeians, and the economic and military requirements of having to rule a far-flung domain. During various twists and turns of the story—from the abortive reforms of the Gracchi brothers, to the military rule of Marius and Sulla, to the defiance of the senate by Julius Caesar, to the final consolidation of power under Caesar's nephew, Octavian, after 31 B.C.—the republic became the empire. Octavian was, in republican terminology, "the first citizen" (*princeps*). But the senate also conferred on him the honorific and semireligious title "Augustus." At first, Octavian may have been reluctant to claim the title "emperor," but he was what we now call Emperor Augustus (27 B.C.-A.D. 14). What was possible in the Augustan period was (what social scientists in our time would call) "a rationalization" of Roman society at home and abroad: a regularization of procedures in law, commerce, public works, and imperial administration. By the beginning of the third century A.D., the empire was fairly well united. Moreover, it was administered with a healthy respect for the local customs of conquered peoples, although in the context of an overlay of Roman language, law, and, importantly, citizenship.

Roman cultural achievements are well known and are dealt with adequately in the main textbooks for which this book is a supplement. But a few points about Roman culture should be mentioned here because we shall return to them. The late republican and early imperial age (about 70 B.C. to A.D. 150) saw the production of most of the memorable works of Roman literature (for example, the speeches of Cicero, the war commentaries of Caesar, Vergil's *Aeneid*, Livy's *History*, and, later, Seneca's drama). These and other works raised Roman culture and language near to the level of the Greeks. And, along with the Greek cultural achievements, Roman culture forms the *classical* base for Western civilization, a base to which Western people would try to return about a thousand years later in the Renaissance.

One of the marvelous paradoxes of history is that, during the time of Greco-Roman glory and grandeur, a new religion emerged within the Roman domain. The Christian religion was based on the Jewish religion, and its thought patterns were similar. Yet, despite its humble origins and its ambivalence about "this world," Christianity was to outdo Rome in influence.

The Romans were great engineers and administrators. Paradoxically, it was along the roads the Romans built, in the towns and cities they administered with the stability of law and in the cultural unity Rome provided, that the Christian religion moved and later flourished. In the terms discussed earlier—Greco-Roman *polis* and Hebraic-Christian *tribe*—Christianity was to confront classical culture on all fronts (ideally) but also to find itself "classicized" (really).

In the long oral history and written history of Christian testimony ("tradition"), it is typically believed that Peter and Paul came to Rome some time around the destruction of the Temple in Jerusalem and that they both were martyred in Rome. Whatever latter-day Protestants might think about Peter as "head of the church," his experiences on several scores gave him a primacy among the apostles as Christianity developed: In the Bible, he alone asserts the words of Jesus' messiahship; to him alone does Jesus apply the term *the rock* (Peter and the confession) on which the Church would be built; to him alone comes the vision that allows the gospel to be preached to the Gentiles. Peter *is* special, and for him to go to Rome seems to indicate two things: that after the destruction of the Temple in Jerusalem there was no longer a locale of Jewishness, and that, in view of the Church's call to preach the gospel to all peoples, it made sense to shift the focus of his endeavors to Rome. Paul apparently believed the same thing, and, not long after their appearance in Rome as leaders of the early church, Peter and Paul were killed. So not only did Rome become important to the Christian church because it was the capital of the empire but also because of certain associations with early Christian choices and martyrdoms.

The Experiences of the Early Church

Christianity grew very rapidly, almost too fast for its own organizational good. There was no central authority, or even local authorities, to govern and to discipline the burgeoning congregations of believers. Yet, it was in the very matters of governance and discipline that authority in the church had to emerge. There seems to be a natural history to this: In the analytical, social-scientific language of our own time, we see the church in transition from a movement to an institution. In its movement phase, there was great growth. Later on, that growth's meaning was assessed, organized, and disciplined, and the early church institutionalized. This, of course, did not happen in a vacuum but in response to real events. On the local and regional level, authority vested in bishops seems to have been fairly established by about A.D. 100. To be sure, the bishops had a board of presbyters (elders) in assistance, but the bishop alone had power to discipline and to organize.

The very early church's experiences were rather fluid and comradely. There was little distinction between members, and they believed, even lived, in common. However, in the biblical accounts of the later books of the New Testament, we already see people with increased authority because of the new needs of the developing organization. As the faith was attacked or corrupted, someone had to assume authority to define what was right. There was, for example, the heresy of Gnosticism (from the Greek word *gnosis*, or "knowledge"). Gnosticism appealed to many people in the churches of Asia Minor, where many converts were either Greeks or influenced by Hellenic thought. The Greek tradition, from Plato and others, valued thought and ideas over "real" matters and the body, and this came into Christianity. Greek thought was a threat to the early church precisely because it was not an organized counterforce but rather a way of thinking within the faith. The Greeks emphasized the spiritual over the material and the hidden meanings of Bible knowledge, if one only could learn them rightly. Although the orthodox were able to condemn this as heresy, this way of thinking within the faith has had enormous staying power in the church, right up to the present. But Gnosticism as heresy was only identified as such through the courageous and tireless efforts of leaders such as Irenaeus, a bishop in France, or Tertullian, a lawyer in North Africa. Most importantly, the Gnostic threat caused bishops to draw up writings from the apostolic ages that were to be read among the faithful. To confuse conditions even worse, the Gnostics claimed special status for certain writings supposedly from the apostolic period.

Wandering evangelists were also a problem. As Paul's epistle to Titus explains, the early church had to be on guard against wandering prophets who were "money-grabbers." They would come to town, ask to preach in the church, and then expect a "love offering." Since these evangelists often made extravagant claims about their experiences and powers, they could often fleece the faithful of their money and go on to the next assembly, leaving the local church bereft of the funds needed to carry on its own ministries.

Indeed, these wandering evangelists often claimed a special charisma (gift), some of which were surely spurious. Many early Christians were very disappointed to find out that those who claimed a special charisma were, as Paul writes, "empty talkers and deceivers," which forced the church to have to organize itself better against enemies from within.

As to the actual history of things within the church, the New Testament writings (whether or not written later) do not take us past A.D. 70. There are, however, church (extrabiblical) documents that suggest that by the year 100, a system of local control by a bishop was nearly universally in place. While in some churches the elders (presbyters) elected the bishop and in other churches the entire membership did, they were still honored as the direct spiritual descendants of the apostles, and their authority was obeyed as if the apostles themselves exercised it. The sacraments of the church were valid only when presided over by the bishop or someone personally appointed by him. He interviewed the traveling teachers, prophets, and evangelists to test their orthodoxy before allowing them to speak in his sphere of responsibility.

The Canon and the Christ

The institutionalization of the church, already seen in the affairs of the local church, comes into even greater focus if we now look at matters that could only be settled in a larger frame of reference: determining what writings were authoritative and the vital question of the nature of Jesus Christ. On the matter of writings, we have already noted that many local bishops had to struggle with Gnosticism, i.e., with determining what documents from the apostolic period were authentic and which of them agreed with the overall body of doctrine emergent among the orthodox faithful.

It was not always easy to determine the authenticity of writings because certain documents, letters, and "gospels" came from the apostolic period and seemed to be of apostolic origin. For example, the "Gospel of Thomas" and the "Preaching of Peter" were apparently contemporary testimonies that had wide readership and acceptance among certain churches. The question was, if the writings of Matthew, Mark, Luke, and John were authoritative, why not those of Thomas and—above all—Peter? Then there were the writings of eminent church leaders of unquestioned orthodoxy, for example, the letters to the churches by Ignatius of Antioch or the letter of Polycarp to Philippi. The question was, if the letters of Paul—not an apostle in the conventional definition—could be given equal weight to those of the apostles, why not those of clearly gifted leaders like Ignatius and Polycarp? Then, furthermore, there was the question of the "correct" version of acceptable writings that may have been embellished in the retelling or recopying. For example, some of the Gnostic itinerant preachers had versions of "the gospel" that gave them special credence for their "teaching," but those versions did not necessarily always agree with other versions (Did Jesus strike some people dead? Did Paul baptize lions in the desert? Did John drive bugs out of an inn so he could rest on a missionary journey?). The Gnostics, among others, claimed to know special truths. The question was, who was to tell which of these versions were authentic and which spurious? And by what authority did they make such judgments?

We must remember that what came to be regarded as the "authentic" Gospels were not written immediately but only in the second generation of the church. The story of Jesus was at first passed by word of mouth. And even when Gospel writers did begin their writing, they did not intend to write a Christian Bible. They already had "the Scriptures," i.e., what we now call "the Old Testament," or the Jewish Scriptures. The Gospels were written to fill a need of telling the story of Jesus to the generations to follow the death of the apostles. But, as indicated above, there were many problems of authorship and version. In the end, and very slowly, a consensus emerged among the leaders of the various churches about authoritative writings.

In this process they used several rules of thumb. The writings, or versions thereof, had to have wide acceptance and broad use throughout the church. They could not contain materials from one region of the church or be the possession of one evangelist, prophet, or group. They had to follow the traditional teachings of the church (although this was difficult because the churches' traditional teachings were emergent rather than defined and only finally codified in general councils in the fourth and fifth centuries). The writings had to be written by, or authorized by, an apostle. Again, this was difficult because some manuscripts were of composite or multiple authority, some derivative of others, some not by apostles (Paul), and one of unknown origin (Hebrews), which was clearly the writing of a person or persons influenced by Hellenism. Notwithstanding the difficulties, a consensus *did* emerge under the leadership of the bishops of the five main churches: Alexandria, Antioch, Jerusalem, Constantinople, and Rome. There were several general meetings (councils) on the subject, and by the end of the fourth century, after councils in Laodicea and especially Carthage, the canon, as later Christians called it, was established. As to the question of the warrant for their authority in establishing an authoritative canon, the bishops claimed that, by the same leading of the Holy Spirit that guided the writers, they were similarly led to accept the thirty-nine books of the Jewish Scriptures and to add the twenty-seven books of the Christian tradition, and to declare them to be, for all time thereafter, "the Bible."

This is a particularly important point to grasp, especially for Protestant readers of this book. "The Bible," as it has been known by Christians since the fourth century, cannot be seen as apart from the history of the emergent tradition of the church. In short, the Bible is what the church said it was. This presents a real difficulty—to which we shall return in a later chapter—for Protestants, who asserted the Bible as "the only authoritative guide for faith and life" and who would then use that same Bible to evaluate the church. In taking the Bible out of the history of the church and using it ahistorically, the Protestants stand at marked variance with the church, which declared the Bible to be the canon. The growth of the church authority must be seen as a response to the need to make decisions about the content of the Christian faith.

The nature of Jesus Christ—a matter of primary importance to Christianity—was a matter needing to be resolved. In the process of defining his nature, the rise of institutional authority is quite evident. Arius, a prominent pastor in Alexandria in the late third century, advanced a doctrine (known as Arianism) that insisted on a radical monotheism. God alone was God, he said, and to assert anything else is to dilute the nature of God. Jesus was indeed the "son of God," but he was created by God at a later time (not exactly from the beginning of time) to show us the way to God. And, the Holy Spirit guides believers in another mode after the time of Jesus on earth. But, Arians insisted, in no substantial sense could Jesus and the Spirit be considered "God" or equal with God. This teaching received wide acceptance, both in Egypt and in various parts of the empire to which Christian missionaries of Arian persuasion had gone. The bishop Athanasius applied discipline to this doctrine by formulating a trinitarian creed. The problem was that the earliest creed, from the apostolic period, the so-called Apostle's Creed, was short, and it admitted to several plausible readings on the question of the nature of Jesus.

In honesty and truth, the doctrine of the trinity is difficult to explain or prove conclusively. But the point is that in two general councils of the churches, first in Nicea (325) and again in Constantinople (381), the whole church asserted and defended "the mystery" of "the three in one, and one in three," declaring it to be authentic and authoritative Christian belief. In doing so the church also condemned as heresy contrary doctrines and as heretics those who held them. In short—like the Bible—Christian doctrine is what the church said it was at a given time and place. The creed formulated at Nicea (the Nicean Creed) is, forever and for all time, *the* standard of Christian belief. Again, one cannot separate the church from belief. In both of these cases—on Scripture and on doctrine—we see a development in the church of authority vested in bishops and councils. The bishops in main cities of the empire were "first among equals" with the other bishops. And, over time, the bishop of Rome was increasingly looked to for leadership in the church, roughly paralleling the secular authority of Rome in the Augustan period.

The churchly and secular authority would converge and then merge early in the fourth century: in a phrase, the "New Testament church" in the West became the "Catholic church" as the Emperor Constantine assumed the dual role of emperor and guardian of the faith (*rex et sacerdos*). But before coming directly to Constantine it is well to say a few words about the relationship of the empire to Christianity. Until A.D. 64 and the reign of Nero, Christians were treated fairly well, both because of the general toleration of local customs in the empire and because, if Rome noticed the relatively small Christian sect at all, it was seen as a sect of Judaism, which had a historic place of protection in the empire. But toward the end of the first century, persecutions began for at least two reasons: The numerous Christian converts in Asia Minor were not Jews, and they tended to be the poorer, lower element of society. Thus, in both cases, because they did not enjoy the historic protection afforded Jews, they became visible targets for those in the empire promoting the worship of the emperor. Persecution, pursued fitfully if savagely under Nero, grew under Trajan (A.D. 112 and after) and increased throughout the third century, when Christians were even stripped of citizenship under Decius and Diocletian. It was in the early fourth century that the dramatic reversal took place under Constantine.

The Constantinian Legacy

Historian Frank C. Roberts (1981) rightly states, "Between the years 303 and 325 changes occurred that are among the most important and spectacular in all church history. From a detested, illegal religious body, seemingly on the brink of defeat, the church became a victorious organization, vehicle of the favored religion of the empire." In 312, and on the eve of a battle that would result in control of the western half of the empire for the winner, Constantine turned from his pagan gods and prayed to the Christian God. The Roman historian Eusebius records the events as follows:

While he was thus praying . . . a marvelous sign appeared to him from heaven. He saw with his own eyes the trophy of a cross of light in the heavens, above the sun, and bearing the inscription "Conquer by this."

Constantine placed the sign of the cross on the shields of his soldiers and was victorious in battle, which convinced him of the power of the Christian God. He converted shortly thereafter, and the history of both empire and church were to be radically different from what one might have predicted only a few years before that. Constantine made Christianity a legal religion in the empire (the Edict of Milan, 313). And, after 323 when he consolidated both eastern and western parts of the empire under his control, he further extended the benefits of Roman authority to Christianity. Out of the state treasury he paid for Bibles to be copied and distributed; called a general council of bishops, paying for travel and lodging expense; paid to have churches and shrines built. He went even further: He gave preference to Christians in government appointments; he sought the advice of bishops in secular affairs; he gave the bishop of Rome a substantial palace for his residence. In time, Christianity moved from being *a* legal religion in the empire, to *the state religion* of the empire by 381.

Christians, especially since the Reformation, have disagreed on the significance and worth of Constantine's efforts on the church's behalf. The majority viewpoint has it that Constantine did a great service for the faith because, as a result of his efforts, Christianity could accomplish its goals of a universal proclamation of the gospel. In any case, the majority view holds, without Constantine's protection Christianity may not have survived very successfully. Surely Christian doctrine and the Christian Scriptures never could have been regularized and systematized had not the general councils of the church been called (and financed) under imperial leadership. He gave the church enough temporal strength to resist the later invasions of the pagan "barbarians" from the north and the Muslim "infidels" from the east. The minority viewpoint on Constantinianism among later Christians is that it began an unfortunate tendency of state interference in the church's affairs, especially in depending on the state for moral and financial support. Many believed that the kingdom of Christ is *never* coincident with the kingdoms of this world and that from Constantine's time onward the church, however "successful" in worldly terms, progressively lost its "soul," until, about a thousand years later, it was no longer really the church at all.

As the empire became more Christian, the church became more Roman. As stated above, the transition from sect to church, from movement to institution, would probably have happened anyway because of the natural history of successful movements, which must consolidate and codify their beliefs so that ensuing generations may be taught. In fact, the institutionalization of Christianity, already underway by 100, was accelerated after 312 by Constantine. Whether the empire or the church was better or worse off because of this can be honorably debated on both sides. However, it seems that the two kingdoms begun so differently would never again wholly disjoin themselves.

Chapter 5

THE MIDDLE AGES:
A PROBLEM IN DEFINITION AND
SPIRITUALITY

History is—I hope we are agreed by now—*our* engagement with the past. In discussing the Middle Ages there are certain questions or presuppositions we bring to the dialogue. Modern Christians take for granted certain definitions about the Middle Ages, and they are shocked when they first encounter differing definitions. Roman Catholics—the majority group of Christians—tend to see the Middle Ages as an unbroken thousand years of Christian ascendancy from, roughly, the fall of Rome to the Renaissance. Here was the great age of Christendom, it is said, with great theological and devotional classics written that parallel the building of cathedrals, the great architectural statements of the faith. While there were many necessary reform movements in the latter part of the period, the Middle Ages represents a time of triumph, not tragedy.

Our Questions about the Middle Ages

Protestants tend to view the Middle Ages quite differently. Some Protestants are comfortable with Constantinian Christianity, but they lament the fall from grace the church experienced with its "corruption" in the late Middle Ages. Other Protestants were never able to affirm Constantinianism, so the slide of late medieval Catholicism into worldly corruption does not surprise them. By whatever route they arrive at it, most Protestants generally believe the Middle Ages to be an unfortunate period in which something "went wrong." Catholic readers of this book will need to be patient for awhile, because for the next few pages we will explore the "Protestant" pattern of questioning about the Middle Ages. However, it should be noted that many Catholics—since Vatican II—would share many of these "Protestant" assumptions.

Let us try to lay out carefully the Protestant line of questions, or assumptions, about the Middle Ages. The basic question, largely unstated, but typically expressed on Reformation Sunday is, was the Reformation necessary and correct? On that day many Protestants sing hymns like Luther's "A Mighty Fortress" and hear a sermon that depicts a medieval church fallen from grace (as John Knox called it, "the whore of Babylon," not "the bride of Christ"). So, something went somewhat wrong (Anglicans, Lutherans, Calvinists) or absolutely wrong (Free Church, Baptists, Mennonites, Quakers) when the early church abandoned the truth, it is said. And the reformers in God's time restored and renewed the true church. Protestant Christianity, then, is restorationist religion, i.e., the view that there once was a time when religion was practiced rightly, and that, after a fall from grace, right religion was restored. Therefore, a restorationist ideology would not be very interested in what went on between the good period of religious activity and its restoration. In short, the Middle Ages are precisely that, in the middle of two better periods—the early church and the Reformation. But it is important to point out that Protestants themselves disagreed on this. Those of Anabaptist or Free Church expressions typically wanted to restore what they called "the New Testament church" (the first century), whereas Protestants of Lutheran, Anglican, and Calvinist expressions wanted to restore "the ancient church" (the first six centuries after Christ) in order that orthodox canon and doctrine be established.

Returning to Protestant assumptions, they say, further, that the Middle Ages were also the "Dark Ages." By "dark" they mean several things but most typically that learning and literacy declined almost to the point of death. Few persons could read and then only nobility and clergy. Moreover, the mass was said in an increasingly irrelevant language. The allegation of religious "darkness" in the Middle Ages is all the more pointed for Protestants because the Bible (the "only authoritative guide for faith and life") was not available to the people in their own languages, even if they could read. Is it any wonder, Protestants ask, that the sermon diminished in importance and that the main focus of Christian worship was the (nonverbal) sacrament of the eucharist? Protestants—irreducibly "people of the Book"—have great difficulty conceiving of Christianity being *authentically* practiced if the Bible is unavailable, unread, or undisclosed in sermons. It is only with the revival of reading and learning, Protestants hold, that appropriate vernacular translations were possible and, later on when Bibles were printed and distributed, that the reformation/restoration of the church was possible.

While one does not want to quarrel with those who insist on the Protestant version of the story about the "dark" and "middle" ages, a few points seem clear. Anticipating the argument of the following chapter for a moment, let us recall that the Renaissance occurs before the Reformation and that it was in the Renaissance that the first suggestion of restoration was made. Humanists in the Renaissance wanted to return to the time of wisdom and light in the classical period, before ignorance and darkness covered Europe. In the rediscovery of the wisdom of Greece and Rome, many humanists believed, Europeans might recover a lost heritage. Protestants and humanists alike looked back to a sort of "golden age." Never mind that one looked to "the New Testament church" or "ancient church" and the other to "the glory that was Greece and the grandeur that was Rome." They both believed that at a prior time the truth was better served, and they both believed that that better time could be restored. The Protestants seemed to have borrowed the ideology of restoration from the humanists of the Renaissance. In the same way they borrowed the methodology that asserted that "reality" and "truth" can be known by the mind of humankind, and it can be written about and talked about. As such, the Middle Ages were dark indeed and best gotten over with so that humankind could get on with the business of being fully human. For purposes of the present argument, Protestants might do well to beware such an (albeit unconscious) affinity with the humanists, and they might be willing to listen to a reformulation of the questions that surround our attempt to define the Middle (Dark) Ages.

Another Formulation of Our Questions

What is an alternative formulation to the questions, where did the church go wrong in the Middle Ages and what was dark about the Dark Ages? It is possible to suggest that the movement in Western society from Roman dominion to the "localism" of the medieval period, when the church was a major cultural influence, was one in which a swing from the polis to the tribe occurred. Of course, we cannot apply these concepts too strictly, but they suggest that the movement from Roman Empire to Christendom is a substantial shift in the center of cultural gravity. As imperial control from Rome diminished, the church—the one institution centered on Rome—continued, and even increased, in responsibility and power. It was the great age of the Christian church, when society became *sacral*, to a large degree. By society become sacral we mean that there was an increasingly smaller distinction between society and religious institutions and ideas.

What a restorationist religious ideology often blinds one to is the continuity of the Christian tradition of the medieval period. Indeed, the whole notion of periodization may even be dubious. For example, we might ask, when do the Middle Ages actually begin? A convenient date might be A.D. 476, the fall of Rome. Yet, if such a date is accepted, one must account for the period prior to that. Surely the time from the birth of Jesus till 476 cannot be regarded as that of "the New Testament church." Indeed, if that period is the one to be restored, it includes the development of a kind of Catholicism that some Protestants may reject.

If periodization presents problems, so does an appeal to special persons, for example, Augustine (354–430). At first a pagan and a pursuer of pleasures, his conversion brought into the tradition one of the most noble "heroes of the faith." He contended for orthodoxy against the heretical views of Pelagius and for orthodoxy on "salvation by grace alone," which was upheld at the councils of Carthage (416) and Ephesus (431). But the work for which Augustine is most remembered is his *City of God* (418). The city of Rome, the so-called eternal city, had been sacked by Visigoths under their leader Alaric. Some pagan Romans said this happened because of Christianity's presence in the empire. They alleged that all had been well while they worshipped their own gods, but when they took over a foreign god ("God" to the Christians), Rome was conquered. Even the Christian Romans wondered how they could have been abandoned by God, because they—like the Israelites of old—believed that their God was the "Jehovah, mighty in battle." Augustine answers both concerns in *City of God:* To the pagans he asserts that the supposed "golden age" of pre-Christian Rome was not so golden, as the recollection of Nero will suggest; to the Christians he said that earthly cities and empires will, of course, pass away and that only the kingdom of God will last eternally.

Both for Romans at the time and Christians later on, one hears within Augustine's writing an echo of a duality he may have learned in his classical education. While there are "real" things in this world, this kind of Neoplatonism assumes, the "really real" things are of the other world. This is the kind of "two-storied universe" that many persons, Christians and non-Christians alike, have found appealing. In short, the real kingdom and the real life are found in the higher realm, while we, unfortunately, must lead our mundane lives in the lower realm. That is a Christianity at home within the dualism of classical thinking.

Continuity in "God's World"

What does it mean for us—in the *doing* of history—to suggest continuity rather than discontinuity? What does it mean for us to look away from the sterile notion of "dark" ages standing between two supposedly better periods? In answer, let us try—with empathy—to enter the medieval world.

At the outset it must be clear that whatever medieval culture was, it was built upon, and among, ruins (both actual and metaphorical). The breaking, or at least fraying, of the threads of Roman dominion was nearly complete at the higher levels of political, legal, and economic life. Yet, the church was the one institution that held Europe together, or at least touched all of Europe. In the thousands of disconnected, especially rural parishes, the priest was the main link with the outside world. Because he may have been the only literate and numerate person in the parish, he assumed administrative duties of a secular nature (keeping accounts, drafting documents) that paralleled his special position in spiritual matters. Here is the problem for modern persons in understanding the Middle Ages: Because it had only religious unity—no political, economic, administrative, legal unity—it must have been an age very "dark" indeed.

Let us, however, follow historian Crane Brinton in outlining what he calls the signs of the tastes and flavors of medieval culture. First, there is the idea of the supernatural as natural, i.e., of God in the ordinary and everyday parts of life. God and the supernatural are not relegated to some upper sphere, with human rationality and purpose governing a lower sphere. Christians in our day may still believe in the supernatural, but the boundaries of what we demark as the supernatural's realm have become increasingly smaller. For example, Christians in agricultural areas may pray for rain to save the crops, but they also consult the reports of the national weather service and the agriculture departments of the state universities. Similarly, heaven and hell have been discarded by most moderns, and some Christians even euphemize them into "presence with God" or "the absence of God." Not so for the medieval person, for whom spiritual realities were very present, even on the evil side, which gives the Middle Ages its reputation for "superstition." The age was filled with wonders, of both the positive and the negative kinds. The spiritual battle was "real" reality. Today people wear crosses as pious symbols or as mere jewelry. The medieval person wore a small cross on his or her clothing because there was "real" power in the cross of Christ that could combat the evils that confronted an individual in the normal course of events. This can be seen in literature and architecture: In Dante's *Inferno,* hell is terribly "for real," not merely metaphor, as are the gargoyles atop cathedrals.

The second major characteristic of the medieval style is the acceptance of pain, suffering, violence, and death as the normal condition of humankind. A modern person might protest here and point out that modern warfare and social upheavals make our own time more violent than the period of knights on horseback and armies with axes. Fair enough, but in our time we see violence and death as abnormal. We look at the concentration camps of the last war and lament "man's inhumanity to man." The medieval person—not our moral superior—would have joined us in our sadness at the sight of holocausts but would not have thought it unusual. We, who have grown accustomed to settling disputes by law and to prolonging life by medicine, have also grown accustomed to a degree of human control over events. Permanence is normal, and those things that are out of our control are abnormal. Medieval society was centered on the acceptance of the impermanence of life. For example, when Easter came again in the spring, the priest did not have to labor in his sermon to suggest the full meaning of renewal. The person in church knew that he or she had made it through the winter when a "chill" could have given a fatal illness ("catching their death of cold"). To have survived, for the medieval, was to have lived and been preserved by the presence of God. No wonder their weekly—sometimes daily—custom was to attend mass, though not necessarily to participate in the eucharist ("the thanksgiving").

In modern terms, the medieval person was used to a hard life. But what we mean by a hard life was also understood by the medieval as hard. The difference between us and them is that we expect something to be done about trouble through medicine, law, peace conferences, and "rights," whereas they saw life precisely as out of their hands and in the hands of God or of those whom God appointed and anointed in the "natural" order of nobles and peasants. Marxists may want to insist that this is the religion of quiet acceptance in the hope of a better life elsewhere and is therefore a trick on the ordinary people. However, for the medieval, not only was there hope in a future world, there was hope in the present world because—in all its uncertainty, death, violence, and pain—one could know that God was in control. As Brinton concludes on this point, it was "not a happy, not a contented world, for in such a world men would usurp the place of God. It was, quite simply, God's world."

Evangelizing Europe

In trying to recapture the flavor of the Middle Ages, we do well to recall that Christianity had had its initial success in the Mediterranean world and only later in what we now call Europe. As suggested earlier, Christianity grew within the Roman world even as its empire crumbled. While there were significant numbers of conversions in Europe by about A.D. 500, the peasant population in many parts of Europe remained steadfastly pagan. Monks were the Christians who would change that.

By about 500, the western part of the Roman Empire was in shambles. Roman authority was largely gone, and barbarians were overrunning the formerly Roman world. The church was saved— and Europe was substantially Christianized—through the work of monks and monasteries. Modern Christians find it difficult to enter the monastic world. We must exercise our empathy to understand these exciting Christians.

Some believers, right from the first century onward, tried to (as they said it) "imitate Christ." Jesus, they observed, had nowhere to lay his head. He lived a very simple and frugal life, oftentimes in wilderness settings. He did not marry. Beyond Jesus' example was his teaching: that followers of his should depend wholly upon God and not fret about tomorrow's needs. For many Christians in the early medieval period, to follow Jesus Christ was to significantly reject "worldly" aspirations and satisfactions (readers of this book from the Anabaptist and "holiness" expressions of Protestantism will have less difficulty understanding this than those from "world affirming" expressions).

Several monks among Egyptian Christians stood out in their devotion to God. Anthony (251–356) and Pachomius (285–346) set a high standard of personal holiness. Moreover, they set a high standard of community solidarity that was based on an essential doctrine of Christianity, "the body of Christ," the true fellowship of believers. As the monks gathered together, they developed rules to govern their common lives.

Benedict of Nursia (480–542) was a most important Christian in the early medieval period in his successful attempt to write rules for the functioning of common life. Benedict was born into an Italian Christian family of considerable worldly means. His rejection of worldly satisfactions and ambitions resulted in his establishing a community near the Italian town of Monte Cassino. Monks were "brothers" in the family of God, and they lived an obedient and frugal life together under an abbot, the community's leader. The monks worked and prayed together and in all ways developed a communal spirit. Benedict wrote down rules that governed communal life at Monte Cassino, and those rules became one of the main standards of European monasticism. Benedict believed that reading and studying were important. One of the main benefits of the monasteries to society was that they kept learning alive in an age when few people—pagan or Christian—cared about such things.

Evangelization was another great impact of the monasteries. A major difference between monasticism in the Mediterranean area and in Europe was that, in the former, there was a substantial Christian population from which the monks drew support and for which they were heroes of the faith. But in western Europe, and all the more so in the north, the countryside was populated by a peasantry largely untouched by Christianity. In Europe, then, a normal part of the monasteries' religious duties was the evangelization of the peasantry.

A model for the new style of monks came from an unexpected part of Europe: Ireland and Britain. It was unexpected because such geographically fringe districts could not have been counted upon to exercise the leading role they in fact did. By virtue of their undaunted missionary efforts, Irish and British monks were to have an important impact on medieval Christianity.

Patrick of Ireland, the most successful monk-missionary, is well and justly known. Born in western Britain in the late fourth century, he was enslaved by pagan Irish raiders, and he lived among them for several years. He later escaped, went to Gaul (France), and learned of the potential for evangelism in monastic dedication. Patrick returned to Ireland in about 432, and by the time of his death in 461, Ireland was not only predominantly Christian, it was strong enough in the faith to send missionary monks to the Gauls and especially to the still largely pagan Germanic peoples. In this regard, Columba (521–97) and Columbanas (530–615) also became legendary in their successes in places outside Ireland, ranging from Scotland to Burgundy, in France.

British monk-missionaries came to the fore a bit later than their Irish counterparts but with no less impact. After the evangelistic success of Augustine (the missionary, not to be confused with the bishop of Hippo) early in the seventh century, and the outstanding organizational work done by Theodore later in the century, the English church was strong enough to send monk-evangelists to

THE MIDDLE AGES / 57

central Europe. The most famous of these was Boniface (680-755), a Benedictine who has been called "the Apostle to the Germans." Boniface combined missionary skill and organizational genius. Church authorities in Rome became aware of his successes, and in 722, Pope Gregory II recalled him to Rome, consecrated the monk as a bishop, and sent him back to Germany. As bishop of Mainz, Boniface was to have a singular impact on the development of Christianity among the Germanic peoples. Because of his British background and connections, he was able to bring a steady stream of Irish and British monks over to Germany. The latter were instrumental in founding many centers of evangelism and learning throughout Europe. Even though the Irish monks were initially the exponents of a localist, "Celtic" Christianity outside Roman control, after the Synod of Whitby (663), the growth and consolidation of Christianity in Europe was not a faith of generic or localistic kind but of a Roman kind. Under the outstanding pope Gregory I (596-604), the Benedictine order became a galvanizing force for the evangelization of Europe and for consolidating the gains under the organizational aegis of the papacy, in this case, the former Benedictine brother, Gregory. Historians Sidney Painter and Brian Tierney (1970) conclude correctly,

the Irish and English monks made Europe a Christian continent by converting the humble, primitive, peasant folk of the countryside to their faith. The work of St. Boniface in particular influenced the development of the Western church in two ways. His missionary labors brought a great part of the Germanic peoples in the framework of a Christian Europe for the first time. And, his activities in Gaul established closer ties than had ever existed before between the Frankish Church and the papacy.

The world of the Middle Ages in Europe was a Christian world. But it would be a mistake to characterize that world as one solely based on the quietistic acceptance of God's will. While nearly everything went on within a consciousness of the purposefulness of God, there was much human purposing as well. For example, the rise of the universities was part of the Europewide struggle to deal with the reintroduction of classical thinking. Also, the building of

Europe's cathedrals was an index of shifting emphases in medieval Christian thinking. Let us take these examples in turn.

Apologists and Scholastics

We must distinguish between two groups of Christian thinkers, separated by time and temperament, the Apologists and the Scholastics. The Apologists lived early in the Christian era (the second to fifth centuries), and they made a defense of the new faith in the context of classical culture. In doing so, they used classical forms of reasoning. What united the work of such diverse thinkers as Justin (Rome), Origen, and Clement (Alexandria) was their belief in the unity of all truth. In their views, whatever pagans like Socrates, Plato, and Aristotle said rightly, Christians should also affirm because all truth is God's truth. Revelation, the Apologists thought, would enable Christians to know more than pagans, in the unity of God's truth.

The Scholastics received their name because their work went on, largely, in universities or schools; they were often called "the schoolmen." The task for Christian thinking had changed by the time of the Scholastics (say, 1150 onwards) because the context had changed. In the second millenium A.D., no one felt the need to give a defense of the faith in a Europe already substantially Christian. God's truth was assumed to be both universal and true. As Anselm (of Normandy and later of Canterbury) said, the relationship of belief and knowledge was assumed. The task of Christian thinking, therefore, was not to give a defense for the faith in a hostile world (like the Apologists) but to advance knowledge in the context of Christian belief. In a largely Christianized world, they sought to *do* theology ("faith seeking understanding").

Thomas Aquinas—along with Anselm, Peter Abelard, and Peter Lombard—best represents scholastic thinking. The task as the Scholastics saw it was to give a Christian account of pagan writings, i.e., "What hath Athens to do with Jerusalem?" It should be mentioned that interest in Aristotle, as well as most of the Greek thinkers, had largely died out in the West after the fall of Rome. However, within Islam there was a continued interest in Greek writers, especially Aristotle. Christian and Islamic culture confronted each other and, to a certain extent, commingled, in medieval Spain and Sicily. Aristotle's writings came back into the mainstream of Western thinking through the "back door" of the Christian listening-in to the way in which Islamic scholars wrestled with the questions of faith and reason. More than any other Islamic scholar, Ibn Rushd (1126-98)—his Latin name was Averroes—is responsible for bringing back into European religious thinking the problem of reconciling reason to faith. When translations of Aristotle's works were completed (from Arabic to Latin) early in the thirteenth century, *the* central question became, if reason contradicts faith, then what? For example, Aristotle demonstrated through reason that the personal immortality of the soul was impossible, whereas Christianity teaches the eternal survival of the individual soul.

Let us be clear about the high stakes involved in the wrestling with Aristotle. On the one hand, a group of thinkers represented by Bonaventura (1217-74), the minister-general of the Franciscan order, believed that Christians need not wrestle with the challenge of pagan thought. Emphasizing intuitive, as opposed to rational, knowledge, Bonaventura and his group believed that the Christian faith opened a person to the revelation of God and, therefore, to all real truth. On the other hand, there was a group that replicated in a Christian context what Averroes had accomplished in an Islamic one. Siger of Brabant (1240-84) perhaps represents this group in believing in a "double truth": that a person might know one truth as a Christian and another, contradictory truth as a philosopher.

Thomas Aquinas (1225-74) was the greatest thinker of the Middle Ages. An Italian of noble birth, Aquinas was a priest in the Dominican order, and he did his famous work while teaching at one of the centers of European thought, the University of Paris. We stop to consider Aquinas because his most important writing, the *Summa Theologia,* may be the most important theological work ever written and because his work lays the foundation for the still-official teaching of the Roman Catholic church. Aquinas believed, like all orthodox Christian theologians, that the Scriptures and the creeds of the general councils of the church were true and authoritative. Aquinas also believed that reason and faith were mutually supportive in the whole life of a Christian scholar. To deny the questions of reason, he said, would leave a believer with an unexamined faith built on assertions. And, in the realm of the natural, God expected us to use the rationality inherent in our humanness. But life is about affirmations, too, Aquinas said, and in the realm of grace we can and do know certain things about God and ourselves that "natural man" cannot know. The questions and the affirmations, Aquinas insisted, did not contradict each other but operated dialectically. While, in due time, some Protestants would make much of Aquinas's alleged splitting reality into the duality of "nature and grace," it is enough for now to say that *Thomism* is the flower of medieval thinking and its symmetry and unity are at one with an age that tried to see, and to live in, the unity of God's world.

Architecture as Index of Change

In the building of the great cathedrals and churches we see both the unity of Christendom and the development within Christian thinking that may help us return to the questions with which this chapter began. The major architectural statements of medieval life—the cathedrals—demonstrate the centrality of Christian affirmation. Moreover, the Christian character of medieval architecture is at one with the other cultural artifacts of the period, i.e., a synthesis of all other aspects of life into the religious. While cathedral-building illustrates the "Christian" character of the Middle Ages, all cathedrals were not the same. We can discern some important changes over time. While there are several variations, two types of church architecture—Romanesque and Gothic—

help us to see the several faces of medieval Christianity. Over the next few pages we will inquire into the differences by looking at two churches in France, Mont-Saint-Michel and Chartres.

Mont-Saint-Michel is Romanesque and Chartres is Gothic. In the transition from one form of architecture to the other, we can perhaps see the developments in Christian "mentality" in the Middle Ages. Mont-Saint-Michel shows the story of the salvation of the elect and the damnation of the wicked. In sculpture and painting, events and scenes from the Bible portray the sufferings of Christ and of his plan for redemption. Chartres is less heavy and solid in outward appearance, and it features the work of the Virgin Mary, whose love and gentle care for all are portrayed. In Romanesque, the observer is awed by the sufferings of Christ. In Gothic, one is reassured of the superintendence, for the good of all, by the blessed Virgin.

Mont-Saint-Michel can be compared to *La Chanson de Roland* ("The Song of Roland"), an epic poem representative of, and evoking, one strand of medieval Christendom. In the poem, we meet the warrior, Roland, dutifully carrying out his mission to his feudal lord. In the pursuit of this duty, Roland dies. In the cathedral, this mirrors Christ's laboring and suffering at the command of God the Father, the ever-faithful Lord, never false to his word. The reconciliation of faith and reason is at one with the Romanesque spirit, which saw what the early church and Scholasticism saw, a possibility of reconciling Greco-Roman thinking and Christian believing. The God of the steadfast word is shown in massive and solid architecture, reflected in a philosophy predicated on an ultimately coherent universe.

Chartres cathedral was built later than Mont-Saint-Michel, and we see more emphasis on mystery and awe than we had seen at Mont-Saint-Michel. Chartres is Gothic, and it is irreducibly Christian, but a Christianity presented in a different way. Dedicated to the Virgin, its architecture, sculpture, and painting are oriented toward a loving Mother who intervenes in the world to protect the helpless and the poor, expressing in lovingkindness the mystical union of all creatures great and small. In literature this is mirrored in the so-called miracle plays (*Les Miracles de Notre Dame*) in which Christians rely on the Virgin for deliverance from trouble. We no longer see the obedient vassal, as in *La Chanson*, but rather merit inhering alone in the Virgin. The Trinity is not emphasized at Chartres, but the unity of the Mother and the Son. *Les Miracles* portrays the Virgin as "the only hope" for the poor in spirit. In philosophy, Chartres is further mirrored in the mystical works of Bernard of Clairveaux and of Bonaventura, insisting on the total brother-and-sisterhood of all persons and all living things. At Chartres we do not see a Father and Son who suffer but a Mother who is love. Here we do not see a sovereign God damning sinners but a loving Mother guarding the helpless. Just as Mont-Saint-Michel can stand steadfastly without support because God has made a balanced and ordered world, Chartres needs the support of flying buttresses to hold up its walls, i.e., the faith that holds the world together. While Aquinas's *Summa* was Romanesque, in reconciling faith and reason, it was also Gothic in its desire to build grand edifices of faith. Gothic also believed in the unity of truth, but it wanted to inspire awe and faith more mystically.

Both Mont-Saint-Michel and Chartres are Christian buildings—of that there can be no doubt—but they are buildings representing different strands of the medieval worldview. While the dates of their construction are not too far apart, the one tends to reflect the earlier medieval period, when there was a greater emphasis on biblical Christocentrism. The latter reflects later medievalism, when there was a greater emphasis on Mary and on a theology of "works righteousness" in which the Virgin and the saints worked on behalf of ordinary people's souls. By now, Protestant readers of this book can perhaps see anew the pattern of disquiet they felt about the Middle Ages, i.e., the vague but deeply felt belief that something had "gone wrong." By now, also, they will, one hopes, have come to see anew that "gone wrong" and "stayed right" may not be particularly helpful terms. In any case, liberated as we are in this book from the requirement of assigning praise and blame, we can see anew what the Middle Ages were, and what they were not, in the approximate millennium between the fall of Rome and the Renaissance.

Something *had* happened, though, in that long period. While there was an overarching unity given by the church, it was not a period of a single, structural unity. During this time, the population of Europe doubled (at least), but in six short years (1348–54) a plague, the Black Death, reduced that population by about one-third. By 1300, moreover, the larger population and the revival of trade created new patterns of social and economic life, all of which contributed to the loosening of the ties that bound serf to lord and penitent to priest. It was in the towns that the revival of learning and literacy paralleled the revival of trade. Many of the towns, begun by the Romans, were reviving, and it should be small wonder that ideas and behaviors more akin to the Romans also revived. Even on the land, the so-called feudal system, which had survived (if imperfectly) for centuries, was breaking down in the face of an increasingly "market economy." As historian Sidney Painter wrote in 1970, "By 1300, the old, personal relations between lord and vassals had disappeared to a great extent. What was left was largely a set of financial obligations." If there was a passing of the old way of life by the fourteenth century, there was also the beginning of a new one. As in all major turnings, there was much continuity along with the change. But we sense a spirit different from the Middle Ages in the Renaissance, to which we now turn.

Chapter 6

THE RENAISSANCE:
THE REVIVAL OF CLASSICISM

In beginning of a discussion of the Renaissance we must be aware that we enter one of the great battlegrounds of the debate about the history of Western civilization. The battle of words rages because the issues that cluster around "Renaissance humanism" were—and are—of great importance, not only in getting the story of Western civilization right but in assessing the meaning of the story in our time. Whether or not the battle is between secular or Christian interpretations, or between several interpretations within a Christian viewpoint, the stakes are high in interpreting the movement (Is it "progress" or "regress"?) from the Christendom of the Middle Ages to the secularity of the modern world.

The Naming of the Renaissance

The agenda for the debate was set by historian Jacob Burckhardt in his famous study *Civilization of the Renaissance in Italy,* published in 1860. This book has had an enormous influence. Many observers—Christians and non-Christians alike—have begun their interpretations with Burckhardt, so we must, too. So rooted in our historical imagination is the name "the Renaissance" that it is astonishing to learn that the name itself was probably first used in the mid-nineteenth century by a contemporary of Burckhardt, Jules Michelet, who wrote the now-famous phrase "the discovery of the world, the discovery of man." In another place, Michelet writes, almost breathlessly, that the Renaissance "runs from Columbus to Copernicus, from Copernicus to Galileo, from the discovery of the earth to that of the heavens. That is when man found himself." Burckhardt takes this line of argument even further. The revival of antiquity, he insists (especially in Italy, with its living memory of Roman grandeur), marked the beginning of a different period of Western history. At marked contrast to the "other-worldliness" of medieval Christendom, the Renaissance brought in a secular, worldly spirit. In contrast to the corporate and resigned social relations of the Middle Ages, the Renaissance was to celebrate an assertive individual as its ideal human type. Burckhardt particularly stressed that the mold of naive faiths was broken by the Renaissance humanists and that, after them, there was no way back to the God ("god") of the innocent Middle Ages.

Many writers have followed Burckhardt's lead, although they may take his conclusions in opposite interpretive constructions. Historians and other writers who celebrate the tradition of the Greco-Roman ideal (Protagoras: "Man is the measure of all things") in turn celebrate the rebirth of that ideal. The whole notion of "rebirth" is itself an affirmation of that tradition of thought, a tradition, it is said, brought to full fruition in the eighteenth century Enlightenment, when humankind becomes fully "modern." As noted earlier in our chapter on the Middle Ages, this strand of Renaissance thinking sees the millennium prior to the Renaissance as an unfortunate one and better for humankind to be done with. Another contemporary of Burckhardt, John A. Symonds, writing in 1888, makes the contrast as stark as imaginable: "The mental condition of the Middle Ages was one of ignorant prostration before the idols of the Church—dogma and authority and scholasticism." In the Renaissance, asserts Symonds, there was "an effort of humanity for which at length the time had come, and in the onward progress of which we still participate. The history of the Renaissance is the history of the attainment of self-conscious freedom by the human spirit."

Paradoxically, certain Christian writers have fully accepted the views of Burckhardt and others. But rather than seeing the Renaissance as the rebirth of what is best in Western civilization, they see it as the renaissance of the worst. Several writers in our time see the same Renaissance world as did Burckhardt but, at every turn, reckon to be bad what Burckhardt thought good. For them, the humanism of the Renaissance is wrong precisely because it is pagan, secular, and individualistic. For them, to make an idol out of humankind is to take Western civilization exactly in the wrong direction, which, these writers hold, was to be remedied by the Reformation.

What unites the two views, though, is the uncritical acceptance of Burckhardt's ideas, that the Renaissance is a momentous break with the Middle Ages and that in the reviving of the classical spirit the "age of faith" was ended. Both interpretations share an ahistorical definition of humanism. In short, if you have seen humanism at one time you have seen it at all times. In both cases, they accept that one's entire interpretation of Western civilization is at stake in making a determination about the Renaissance. We should not accept such an easy dichotomy. Following the lead of recent scholarship, it is no longer acceptable to see a sharp break between the Middle Ages and the Renaissance, both because of research done on the Renaissance "spirit" (now accepted to have been present in the twelfth century) and on the medieval "ethos" (now accepted to have continued through to the sixteenth century, in the desire of many Protestants to maintain a sacral society). So, in the remainder of this chapter, we will back away from the "Burckhardt thesis," insisting that it is a mistake to draw such a sharp distinction between the medieval and Renaissance periods, seeing medieval art as about religion and inhibition and Renaissance art as about secularity and exhibition. The *real* break between religion and secularity happens in the eighteenth century Enlightenment, based on the ideological use of science. *That* is the beginning of modern times. In the period of the Renaissance-Reformation—for they must be linked—we see a *transition* period, retaining much of medieval Christianity while anticipating, but clearly not reaching, modern secularism.

Let us now, with historical empathy, try to see how it was plausible for the people of the Renaissance themselves, and their later interpreters, to believe that they were doing a new thing in Western civilization. They believed, and they were right in believing, that their age was one of "discovery." If we lengthen and broaden our definition of discovery, we have a useful tool for organizing our thinking about the Renaissance, because what happened after (approximately) 1350 was a journey both inward and outward for humankind in Western civilization. At the same time there was an expansion of "space," both geographical and metaphysical. As Europeans expanded the definitions of the known world in geographical space, there was a corresponding expansion of space for the mind and spirit.

Discovering the World

The Portuguese were in the forefront of discovery, with motives combining acquisitiveness and altruism. Trade with the East had languished at the end of the fourteenth century, both because of the Black Death in Europe and because the Ming dynasty in China was discouraging external trade. The Portuguese believed they could reestablish trade with the East by going around Africa. Their base for operations was the city of Ceuta in northern Morocco, which they seized in 1415. From there they pressed down the coast of Africa with the support of the royal family of Portugal, most notably Prince Henry (1394–1460), whom the English later called "the Navigator." In 1488, under Dias, the first ship rounded the tip of southern Africa, and by 1499 DaGama had reached India and established trade arrangements with the Hindu rulers. The Portuguese also believed they might convert Muslims in the process of their explorations. Some explorers even believed they might discover the "kingdom" of "Prester John," a Christian who was believed to rule in either East Africa or southwest Asia.

Important as the Portuguese were in beginning the European outward thrust, the later work of the Spanish was of greater importance. With the financial support of Isabella of Castile, Christopher Columbus sailed west to find "the Indies." According to his calculations, they lay about 2,400 miles west of the Canary Islands. As is well known, Columbus arrived at islands in the Caribbean Sea in 1492. He could tell this was not Japan or China, and the physical characteristics of the inhabitants resembled the natives of India. So he assumed that India and Asia were near and, therefore, called the natives "Indians."

The "discovery" of America by Columbus was, at first, a major disappointment for Europeans. They were trying to get to Asia, and the existence of a large continent between Europe and Asia was as disheartening as unexpected. As a later explorer said, "America was discovered when not looked-for, and explored in an attempt to get around or through it." Early in the sixteenth century, however, after many voyages by several exploration teams, it was now clear that this large continent was, in fact, there, and that Columbus had been woefully mistaken in his calculations of the westward distance between Europe and Asia. But, what first was a disappointment soon turned to possibility. Europeans had long wondered about the legend of "Atlantis," a continent of wealth and possibilities lying southwest of Europe in the great ocean. Southern Europeans were unaware of Celtic and Viking exploration and settlements in what we now call North America, but rumors persisted in the south about Atlantis. Now it seemed to be true.

The meaning of the existence of a new continent took some time to develop in the European imagination. Yet, only a few years after the initial discoveries, we see the first major reflection in the literature of the "old world" on the existence of the "new world." One of the masterpieces of Renaissance literature is Thomas More's *Utopia* (1516), a work that literary historians say may well rival Shakespeare's plays as the most widely read literary effort in the history of the English language. More was an inveterate punster, and his title can mean that the new world is either "the perfect place" or "no place." The narrator of the story—a traveler who allegedly went on one of Amerigo Vespucci's voyages and stayed behind—is Raphael Hythloday. In some translations, Raphael is called "Nonsenso," or the teller of nonsensical tales.

More's *Utopia,* then, lays down two lines of thought about the new world that exist to the present in the European mind (and to which we shall return later in this book), that "America" is both/either a paradise or a fool's paradise. Indeed, *Utopia* is not so much about the new world as about seeing it as a mirror for the evils of the old. More, a committed Christian (later, indeed, to be executed for his steadfastness to the faith) never suggested that America was a new "Garden of Eden" that had escaped the Christian Fall but a place where Europeans might remedy some of the consequences of the Fall. The utopians had overcome social injustice by an act of their collective wills, and, holding things in common, they lived at peace and in harmony, with each citizen working for all and all for each.

Returning to our theme of discovery, we can see this Renaissance style in other aspects of culture. Writers began to discover that good literature could be about secular, as well as religious, subjects. Moreover, they could, indeed should, write in the vernacular language of the people. Dante Alighieri, whose *Divine Comedy* is thought to be a crowning work of the medieval worldview, also wrote extensively on the appropriate use of common language (another example of why we should be careful of drawing too fine a distinction between medieval and Renaissance styles). Dante influenced more worldly writers who, in turn, wrote about the experiences of common people. Especially in the writings of the fourteenth-century authors Petrarch, Boccaccio, and Chaucer, we see a zest for living the life of this world and a tolerance for different ways of life.

"Discovery," in the literal sense, also describes the delightful recovery of many "classic" manuscripts presumed lost. One discoverer, Poggio Bracciolini, found an original manuscript by the

Roman rhetorician Quintilian in a monastery. He was aghast that the monks had no sense of the worth of the treasure they had held. Bracciolini wrote to a friend, "We discovered Quintilian, safe and as yet sound, though covered with dust, and filthy with neglect. The books, you must know, were not housed according to their worth, but were lying in a most foul and obscure dungeon at the very bottom of a tower, a place into which condemned criminals would hardly have been thrust." Indeed, the "discovery" of "lost" manuscripts in the West was unintentionally aided by the Turks because, after their conquest of Constantinople in 1453, many Byzantine scholars fled west, bringing with them supposedly lost books. The discovery of these books and the subsequent dissemination of "classic" writings was aided immeasurably by the work of Johannes Gutenberg of Mainz, whose development of movable type allowed books to be printed accurately rather than copied.

The artists of the Renaissance rediscovered the heritage they possessed, especially in Italy, of "naturalistic" art. They acted, in short, on the now-famous dictum of the late medieval thinker Thomas Aquinas (another example of not distinguishing too finely between medieval and Renaissance): "Art is imitation of nature; works of art are successful to the extent that they achieve a likeness of nature." First, perhaps, in Giotto, and surely by the time of Brunelleschi, they mastered the technique of three-dimensionality, and only through that could they liberate themselves from the prior canon of art that was flat and two-dimensional. Merely to list the names of the great painters and sculptors of the fifteenth and sixteenth centuries is to suggest the greatness of the rediscovery of "humanity" in the Renaissance: Donatello, Botticelli, da Vinci, Michelangelo, Raphael, Titian, Durer, Brueghel, Holbein.

In the last few pages we have explored the idea of discovery as uniting the activists of the Renaissance. There seems to be an expansion of space. The willingness to thrust outward from conventional patterns of behavior and belief was animated by a corresponding conviction both about human worth and about the worth of human actions. Emboldened by the recollection of the achievements of Greece and Rome, humanists also came to believe in themselves as well. They attempted to revive that spirit of Greco-Roman striving that had long languished in the West.

In view of the above comments, the perceptive reader will recall, and will want to ask about, our previous discussion about the Burckhardt thesis and about our wanting to back away from antithetical distinctions that see the Middle Ages as religious and the Renaissance as pagan. What may have misled Burckhardt and those who follow him was too exclusive a focus on southern Europe, especially Italy. As the Renaissance moves to northern Europe, we notice the same "exploring-discovery" spirit, but one more sober, tempered, and, in fact, more religious. While one cannot make a simple distinction between the northern and southern renaissances, there are some differences we ought to notice.

Humanism in the Renaissance

The humanists of the Renaissance were rebels against authority, and, in this generic definition, we are not prepared just yet to say what authority they rebelled against. As historian Crane Brinton says well, "This complex movement in the arts and in philosophy we call humanism is a very self-conscious rebel, a rebel against a way of life it finds corrupt, overelaborated, stale, unlovely and untrue."

The best single example of the Northern Renaissance is the life and work of Desiderius Erasmus, who was born in Rotterdam in 1466 and died in Basel in 1533. His legacy is as important as his career is diverse. On the one hand, his *The Praise of Folly* (1509) was written while he was staying at the home of Thomas More in England. Erasmus, a man of letters but not a theologian, was regarded by church authorities as writing on subjects on which he was not qualified and in an improper spirit. One such authority, Professor Martin Dorp of the University of Louvain, criticized publicly the work of his old friend, and it was to Dorp that Erasmus made his now-famous reply. Quoting the Roman writer Horace, Erasmus said that, while he had a serious point to make, "what is the matter in saying truth with a smile?" Erasmus's more "serious" work also made the same point. Especially in what he called "the philosophy of Christ" he sought to reconcile the best of pagan (he would have said "pre-Christian") antiquity (especially Plato and Cicero) to the Christian gospel: "If there are things that belong particularly to Christianity in these ancient writers let us follow them," he said. His pursuit of intellectual honesty led him, most of all, to direct his efforts toward a more accurate version of the New Testament. It was his view—rightly, as we now know—that the translations of the New Testament then available contained textual errors, which sometimes led to theological errors. His work went through several editions between 1516 and 1536, each more exact and annotated. In giving an accurate New Testament to the reading public, Erasmus the humanist (in the delightful epigram of a later wag) "laid the egg which Luther hatched." In a fitting end to the paradoxical career of this prototype of the Northern Renaissance, Erasmus's burial service in Basel was held in a Reformed church, a building that had been a Catholic cathedral at the time of his birth. It is worthy of note that he remained a Catholic.

We can now return to our attempt to define the humanism of the Renaissance. If it be true that humanists were self-conscious rebels against authority, we must make careful distinctions about both "rebellion" and "authority" because the generic definition that includes, say, Boccaccio and Erasmus, claims at once too much and too little. Brinton helps us with some terms that can be of great analytical assistance; he sees two expressions of humanism, which he calls *spare* and *exuberant*. Let us explore the meanings of those terms.

The spare humanists are "historical" in their rebellion against authority. By historical we mean that they rebel against prescriptions about behaving and believing, but they do so in a self-conscious desire to replicate the belief and behavior in an earlier period, "antiquity." For them, rebellion was not so much against all authority, but "wrong" authority, in whose place they would reassert and reestablish "right" authority. In contrast, exuberant humanists, while celebrating the style of antiquity, did not feel obliged in their rebellion against authority to insist that the authority of antiquity is necessarily "better" than contemporary authority. For them, the rule of private conscience and intellectual/artistic honesty (true to one's self) was enough.

The spare humanists are disciplined; they insist that a method of knowing and acting can be gleaned from the "classics" of antiquity. Merely ransacking Greco-Roman culture to prove one's point was not enough. Only careful inquiry would yield truth. The exuberant expression of humanism was, by contrast, "free"; if the human personality is allowed to explore its liberation from dogma, superstition, and oppressive prescription, the truth would emerge. Indeed, there may not even be one truth but multiple truths, to be known as individuals become fully human.

The spare humanists are "corporate" in their rebellion. In their attempt to replace one authority pattern with another (better, in their view), they feel responsible to explain themselves to each other and to a watching world. They are responsible to the collectivity of humankind because it is the betterment of all that they seek. In contrast, the exuberant humanists celebrate the individual. They do not accept collective responsibility or the need for mutually critical collective action. To them, the full flavor of humanity will bloom when, and only when, individuals are liberated from any and all tyrannies over the mind and spirit.

Perhaps the simple graphic in figure 4 will help to punctuate the preceding distinctions.

HUMANISM

Exuberant	*Spare*
Contemporaneous	Historical
Freedom	Discipline
Individual	Corporate

Figure 4.

In using functional terms like these we do not yet say whether or not humanists were "right" or "wrong" to rebel against medieval "authority." It has seemed wiser for historical students to withhold judgment on whether or not it was right (or wrong) to rebel, until we see the manner of the rebellion and the thing rebelled against. Now, in going back to the Burckhardt thesis of a clear disjuncture between medieval Christianity and Renaissance humanism, the one religious and the other neopagan, we see that such a dichotomy does violence to the story. Its insistence on such stark divisions is awry. Yes, the Renaissance *does* see a rebirth of classicism. And, yes, to some humanists the statement of Protagoras *is* accurate ("man the measure of all things"). But to others, the rebirth of classicism did not necessarily mean the elevation of "man" over "God" but a way in which humankind might better know and worship God and live the Christian life in the new context of an age of discovery. Therefore, we can now, at length, come to an approximate definition of Renaissance humanism: Humanism is not so much a *philosophy* as a *method* of questioning authority by which a number of philosophies—both religious and secular—are possible.

All humanists acted on the belief in human potential. But, the question was—and is—potential to do or to know what? Yes, all humanists asserted the worth of humans and the worth of human knowing. But, the question was—and is—humans doing what and knowing what? The context of the Renaissance is tremendously important to recall, that it was Christendom against which humanists rebelled. Thus, on the one hand, much of their "rebellion," especially in northern Europe, was an attempt to *do* Christianity better. Most humanist art was Christian art (Is Michelangelo rebelling against God?), and most humanist scholarship was Christian scholarship (In providing an accurate New Testament, was Erasmus rebelling against God?). On the other hand, there were some humanists whose work, both in style and substance, was an attempt either to replicate Greco-Roman culture or to liberate the free individual. But therein is the point: Humanism is a complex movement, valuing human worth and human knowing. But it is a movement possessed of a methodology that admits to philosophies of life (worldviews) *both sacred and profane.*

In another context, shorn of its religious casing, a new pattern of human assertiveness *will* (necessarily) issue in a secular worldview. But in the Renaissance we do not have that final break. Though the potential for it is definitely there, and will come in time, it does not happen in the fourteenth to sixteenth centuries. What perhaps confused and confounded Burckhardt, and those who follow him, was that humanists used the language of liberty and rebellion. Latter-day students perhaps read the use of that language through the prisms of post-Enlightenment thought and experience. The political side of humanism demonstrates this problem. Renaissance humanists were not forerunners of modern democratic liberals; they were frank elitists. In opposing monarchy—in either the church or the state—they did not propose democracy (although exuberants might have) but a republic, drawing upon the city-states of Greece and the *res publica* of preimperial Rome.

A Christian Humanism?

A final question for this chapter is one that Burckhardt and his intellectual followers (either pro or con) will find difficult: Can Renaissance humanism and Christianity be reconciled? In short, can there be such a thing as "Christian humanism"? But before answering the question with a direct yes or no, let us be clear about why this is a difficult question. The majority church among Christians, the Roman Catholic church, will have the most difficulty with the question. If one accepts, as most Catholics did, that the church was one unbroken testimony of God's revelation—from the day of Pentecost, through the creeds, through the canon of Scripture, and through the history of the tradition—then there can be no reconciling the religion of a God who purposes his will and the self-assertion of mere "man." So, for historical Catholicism, the question really isn't interesting because it turns on the unwanted possibility of humans taking away the authority of God and God's appointees. The question is also not interesting for people who are fully secularized. For them, humanism in the Renaissance began a way of thinking and acting that has culminated in the spirit of democratic liberalism, i.e., that society should be a democracy and that human enterprise, both political and economic, ought to be "free" from controls.

It would seem that the only people with a historic stake in the reconciliation of Christianity and humanism are the people called Protestants (although post-Vatican II Catholics in North America may well be considered as "Protestants" in this regard). Protestants are people for whom the "old-time religion" was not good enough, even if it had been so for their fathers. Protestants are people who interposed their own consciences between themselves and their liberty on the one hand and the authorities of church and state on the other. They rebelled against then contemporary authority in the name of something (they said) "better," and in doing so they conformed to the method of humanism. They say there is wisdom in antique sources (the Bible, "the New Testament church"), which offer a more excellent way than did medieval Christianity.

While it is true that there is a world of difference between citing Plato and Paul as antique sources legitimating rebellion, it is nevertheless true that it is a similar methodology, issuing in a conjunction of concepts that neither the majority of Christians (Catholics) nor the majority of humanists (secular liberals) find persuasive. So to the question "Can there be a Christian humanism?" the answer is, apparently, that most Protestants seem to think so. At this point, an incredulous Protestant reader, perhaps unconsciously influenced by the Burckhardt thesis, may demand an answer to the following question: "Where, then, do we begin our philosophy, with God or with 'Man'?" The answer, in good Christian humanist fashion, at least for most Protestants, is apparently this: "*We* begin *our* philosophy with God." In that simple affirmation lies an entire story about human purposefulness and human worth. It is clear, in that affirmation, *who* is purposing and acting. It is also clear *what,* or better, *whom* they have chosen, and that makes all the difference. In the end, the Renaissance is about the revival of classicism. But classicism was not a single thing. There were certain aspects of the classical heritage at home in Catholicism, as Aquinas showed. But other aspects of classicism's impact—about human worth and the worth of human knowing—were to become more manifest in Protestantism.

Chapter 7

THE REFORMATION: RESTORATION AND/OR FRAGMENTATION

At the beginning of the last chapter we cautioned against making too much of the supposed distinction between the style of medieval Christendom and that of the Renaissance. In this chapter we must be even more careful in seeking differences between the Renaissance and the Reformation. At least the attempt to distinguish the Middle Ages and the Renaissance had chronological discontinuity on its side. The attempt to distinguish the Renaissance from the Reformation does not even have that benefit. Some dates will illustrate the point: More's *Utopia* and Luther's Ninety-five Theses are but a year apart (1516, 1517), and they appeared only a few years after Machiavelli's *The Prince* (1513); the final edition of Erasmus's translation of the New Testament and the first edition of Calvin's *Institutes of the Christian Religion* occur in the same year (1535); the first "Protestant" pamphlet of the former priest-humanist Zwingli and Magellan's circumnavigation of the globe were also in the same year (1522). When we have discovery overlapping so markedly, it is difficult to say when or where the Northern Renaissance became the Reformation or even how we can distinguish between them.

Our Questions About the Reformation

This book sees continuities rather than discontinuities, and, at best, transitions rather than abrupt changes (although we will see a substantial shift in the Enlightenment). But since many Protestant (and some Catholic) readers may well have previously accepted certain conventional affirmations, it is well for us to deal with those first—to clear the air, as it were—before we inquire into the method, nature, and legacy of Protestantism. What are some of the questions surrounding these matters that "everybody knows"?

First is the question asked about "corruption" in the church. It is said that the Catholic church had fallen into a corruption so utter and complete that when the Protestants left they did not leave "the church" but a pretended version of it. Fair enough, one says on this point: There were substantial corruptions in the church, most notably the sale of indulgences in Germany under the infamous Dominican monk and salesman, John Tetzel. In rejecting the sale of indulgences, Luther objected not only to a nefarious practice but to the entire underlying theory of a "treasury of merits" by which certain saints could compensate for the shortfall of righteousness on the part of ordinary sinners. Yet, as important as this is, the question remains as to why a break in the church had to result because of such protest. There had been difficulties, even "errors" in the church before 1517 (as witness the protest of St. Francis of Assisi and his other-worldly followers against the worldly power of the church). But reform movements had mostly, but not always, stayed within the church. Indeed, that is the definition of "reform" (to work within established institutions), because a reformer believes that God establishes constituted authority. That authority may have fallen from its ideals, but one works within institutional structures to call leaders and people back to the ideals.

In answering the question, why did reform become rebellion? we need to look beyond theological rhetoric to the context of the Reformation. It is important to note here that Luther wrote three pamphlets in 1520 that did far more than the Ninety-five Theses to seal the break with Rome. In *The Babylonian Captivity of the Church* he advocated the reduction of the number of sacraments from seven to three (baptism, Lord's Supper, confession). In *On Christian Liberty* he reaffirmed the doctrine of salvation by grace alone and that good works flow from grace alone. In *Address to the*

Nobility of the German Nation he called on princes to use their secular authority to reform the church in the face of the clerical refusal to do so. In the latter pamphlet, written in German not Latin, Luther raised the spirit of nationalism in support of his call to reform the faith. Listen to his rhetoric:

For Rome is the greatest thief and robber that has ever appeared on earth, or ever will. . . . Poor Germans that we are—we have been deceived! We were born to be master, and we have been compelled to bow the head beneath the yoke of our tyrants. . . . It is time the glorious Teutonic people should cease to be the puppet of the Roman pontiff.

The appeal here is clearly for a national church as opposed to the international Roman church. So, when we ask why a break in the church happens after 1517 (since corruption was much worse, say, in 1417), corruption cannot be the main reason, in itself, for the break. Rather, we must look deeper into the context of the Protestant revolt and see the convergence of certain religious ideas with a growing national consciousness.

A second conventional wisdom among a minority of Protestants (but a view fairly widespread) is really a subtype of the first concern about corruption. All Christians believe that the Holy Spirit (the "other comforter") came to the church with fire on Pentecost. Protestants replied to the Catholic charge of "grieving the Holy Spirit" (i.e., if the Holy Spirit is in the church, and if you vilify the church, you oppose the Spirit) by saying that the Holy Spirit had left the church.

This is an important point for some Protestants, so it ought to be discussed briefly. Since, they say, they were restoring real religion to the church, they *must* be supported by the Holy Spirit (who cannot be against himself). Thus, the Spirit *must not* be in the Catholic church. As we can quickly see, this is an assertion that cannot be proved or disproved and is itself a matter of logic, not of fact. It is, however, plausibly a logical fallacy, in which the actor believes that, since *he* is not doing the wrong thing, it must be his opponent who is wrong. This logic becomes fallacious and fragmented when questions of fact are raised. For example, if the

Holy Spirit *did* leave the church, and since you seem to know what others do not know, you should tell us *when* the Spirit left. A mere invocation of "after the New Testament church period" won't do because any orthodox believer "needs" the Holy Spirit still in the church at least until the mid-fifth century, when canon and creeds were firmly established. Context is, once again, important.

A third point "everybody knows" is that Protestants rediscovered "biblical Christianity," i.e., a Christianity that conforms to the Bible. This is an important concern because, surely, Christianity and the Bible are one. Most readers of this book will agree with the serious importance of this point while already having sensed (from a previous chapter) what difficulty here arises, because we must ask the subsidiary question, where does the Bible come from? Apart from the dogma that the Bible (as "God's Word") comes from God, we have noted earlier that the Bible, especially the New Testament, came to be recognized within the tradition of the church. The Bible as we know it was—and is—what the guardians of orthodoxy within the church said it was. Even as we cannot see the Bible emerging apart from Christianity, we cannot see it as apart from the church (at least up through the period of the canon's establishment). Thus, to take the Bible as outside the tradition in which it grew and was established is an extraordinary exercise. Further, to say that the Bible "speaks for itself" is to beg the question of responsibility and authority because it pushes back another step the search for authority. In fact, someone had to read the Bible and give interpretations of it. Even in a very democratic approach of every believer being as "priestly" as any other, someone had to say what the Bible meant, to do the sorts of things the clergy had previously done for them. Here again we see Protestantism as part of its (Renaissance) time, when individuals or groups asserted their own authority over against existing authority and utilized antique texts in that assertion.

A fourth, and last, viewpoint among some Protestants is that the Protestant Reformation provides a "base" from which to judge the past and the future. This view has been widely popularized by the late Francis A. Schaeffer. His widely shared viewpoint runs roughly thus: God's intentions were expressed uniquely in his Word (all Christians accept this), and a unique rediscovery of those intentions occurred in the Reformation (a minority of Christians accept this). What this viewpoint insists upon is taking both the Bible and the Reformation out of the flow of history and making them ahistorical, abstract, repositories of authority. While this may make for satisfying apologetics, it is not acceptable to students of history, who take no movement out of its time. Even if we could do so, the question remains as to which of the branches of Protestantism speaks truly, because, as it is well known, they do not speak with a common voice. It should be noted that the argument for a Reformation base comes typically but not only from the Calvinist expression of Protestantism, an expression in which the present writer stands. But it is one, in intellectual honesty, he cannot say is "normative" because to insist on the normativity of one view among many is to leave aside the historical task and to enter the realm of apologetics, where theological rather than historical points are at issue.

In this attempt to clear the air by giving some notice to some view that "everybody knows," what can be offered by way of conclusion? An ideological Protestant might interrupt at this stage to say that the points offered above, in sum, cast doubt on the "rightness" of the Reformation. After all, he or she might ask, isn't the Reformation "right," and, in spreading doubt about that, whose side are you on? Indeed, these questions are not to be merely dismissed and ought to be taken seriously. (In the introductory chapter, we noted that the *doing* of history requires confrontation with ourselves, so it is OK for a reader to confront the writer.) As to whether or not the Reformation is "right," first let us recall that the historical task is not, in the first instance, about ransacking the past to ladle out praise and blame that can be used for an ideological present purpose. The historical task is about listening to the human conversation. And because history requires *our* engagement, it is also about our entering the conversation. But the goal is always understanding past participants in the conversation (and ourselves), not "setting straight" them or our contemporaries. With a nonjudgmental attitude, liberated from the requirement of praising and blaming, we can then—and only then—learn. Further, if you want to ask me if I think the Reformation is "right," I reply with, "Which Reformation?" Readers of this book may well range among Lutherans, Episcopalians, Presbyterians, Calvinists, Methodists, Baptists, Free Church members, and Mennonites (never mind the various secessions from the mainline denominations, e.g., Free Methodists, Nazarenes, Conservative Baptists, Reformed Presbyterians, and the various "Bible" churches who have "no creed but Christ"). Who among us will say that *our* Reformation was *the* Reformation (never mind the majority of Christians, in the Roman Catholic church, who think the entire Reformation to be an act of disobedience to God)? I realize that some people do precisely that. I once attended a lecture by a person from a small Protestant subdenomination (a secession from a secession from a secession from a major denomination) who insisted that his group alone had "kept the faith" of "the New Testament church" while all others had diverged on some matters of faith or practice. Surely this is an unsupportable assertion, and—as historical judgment—no church or denomination may say that *it alone* is "the one true church." If it does, it ceases being a "church" and becomes a "sect."

Now, since "permission" has been granted to confront the writer, it is OK to ask him the following: "I am a Protestant, and I want to know if you believe that my forebears were wrong in leaving the historic church. What would you have done in their place?" Or: "I am a Catholic, and I want to know if you believe my forebears were wrong in staying within the historic church. What would you have done in their place?" So, it is time to come clean, to abandon the Socratic method of answering a question with another question. I, too, am a Protestant but one who has a few Catholic forebears among the majority of Protestant ones. So I *feel* the weight of the question very deeply. I see the Reformation as a tragic necessity. That is, it was a tragedy because it ripped apart the unity of Christendom, with that ripping sometimes too much animated by the nationalistic and humanistic spirit of the age. And it was necessary because it was a major revitalization of Christianity in an age of increasing literacy, which brought forward a new clarification of Christianity in the various Protestant groups as well as in the historic church, which also clarified its own views in response. So, when Reformation Day returns every October 31, it is for me a day of both triumph and tragedy in which, at once, one celebrates the revitalization of Christianity and one laments the fragmenting of the church contrary to the hope of Jesus, who wanted us who name the Name to be one.

It is hoped that the above, if not answering important questions, has at least acknowledged them and allows us to go ahead—in a spirit of cooperative understanding—to inquire into the nature of Protestantism. In the next few pages, let us try to define Protestantism and see its nature.

The Nature of Protestantism

Protestantism is such a multifaceted movement that the only definition that will fit all parts of the movement is a generic one, such a low common denominator that one wonders about its usefulness. Protestants, it seems, were people who no longer wished to be Catholics. That is all we can say if we want to touch all of them. Because, following the language of liberty, we can know what they together said no to (the historic church), but what they said yes to was so diverse that no single definition will unite, for example, Lutherans and Mennonites.

Within this generic definition, though, we can say some things about the *nature* of Protestantism that may help us when we return to more specific definitions. As Crane Brinton suggests, "Protestantism was a revolt against an established authority possessing the external attributes (organization, laws, ritual, tradition) of authority. It asked men to disbelieve and disobey." So Protestants, like the humanists whose method they shared, were "conscious rebels against authority." Every rebel against authority always uses the language of liberty to legitimate that rebellion. In twentieth-century colloquial language, Protestants are people who had been "had" and they "weren't going to take it any longer." The first act for a Protestant leader, then, is to encourage followers to *disbelieve* in something that they might believe aright and to *disobey* something or someone so that they might obey aright. Herein, then, is the tension within the Protestant movement that causes it to be inherently unstable. The questions are as follows: When rebellion against universal ("catholic") authority is sanctioned by local leaders, how can authority be reconstituted? When rejection of universal belief is sanctioned by leaders in favor of particular beliefs, how can belief be reconstituted? These are real questions, not straw questions to be disposed of quickly; they were—and are—questions that have plagued the Protestant movement since 1517.

Rebellion always uses the language of liberty, but it frequently comes to grief when the anarchy implicit in rebellion becomes explicit. For example, take Luther—surely Protestantism's most courageous initial spokesperson—and his difficulty in reckoning with the "Peasants' Revolt" not long after his own revolt. Luther argued that God would not interpose any "authority" between himself and his creatures. Therefore, let each person be his or her own priest, and read the Bible without priestly intermediary. It astonished Luther, however, that ordinary people might follow his *method* of rebellion but not agree with him on the *object* of that rebellion. He agreed to the (sometimes savage) quelling of sociopolitical rebellion that clearly paralleled his own socioreligious one. As people in the Anabaptist expression of Protestantism will readily attest, Lutherans soon began to interpose authority between God and humankind, but, this time, it was authority of their own making. Rebels against historic authority often wound up recreating a new authority that mirrored the authority originally rebelled against, as will not surprise readers of, say, George Orwell's *Animal Farm*.

Protestantism then—we are still attempting definitions—is a broad and diverse movement that shares a quality of self-conscious rebellion and the consequent structural instability. There is really no *final* repository of authority, because, as noted above, the Bible (rhetorically the only authoritative guide for faith and life) is not, strictly speaking, an "authority," but *becomes* authoritative when interpreted by a person or an institution. And, since many Protestant leaders and institutions saw different prescriptions in the Bible (for example, on baptism), contentiousness is rife within the Protestant movement. Apologetics is a main Protestant concern: Already a minority movement in the whole church, splinter groups become increasingly shrill as their increasing minority status made them ever-conscious of how few of "us" there are and how many of "them."

Perhaps it would be helpful to reproduce in the Protestant connection the distinction between spare and exuberant we first used in the humanist one. In our attempts to define Protestantism, it is best for us to admit that the generic definitions take us only so far and further to admit that it is better to talk about "the Protestantisms."

The strength of using terms like *spare* and *exuberant* in the continuum of the Protestantisms is that they are functional terms and, therefore, need not give offense to anyone. The weakness in using a pair of terms is that there are three "family groupings" of churches, not two. Yet that difficulty is overcome when we recall that these are not airtight categories but points along a continuum, suggestive of distancing from a given starting point of reference.

In figure 5 we see that the historic church is on the far right, as befits its conservative role. Among the Protestants, we see three affinity groups, from right to left. Lutherans and Anglicans (Epis-

THE PROTESTANTISMS

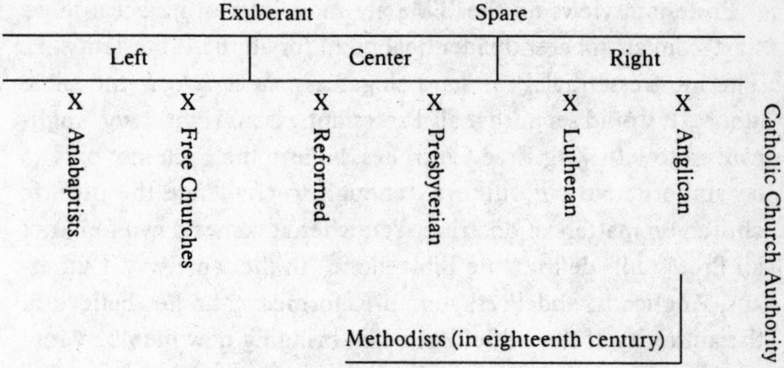

Exuberant				Spare		
Left		Center		Right		
X	X	X	X	X	X	
Anabaptists	Free Churches	Reformed	Presbyterian	Lutheran	Anglican	Catholic Church Authority

Methodists (in eighteenth century)

Figure 5.

copalians in the United States, Anglican church in Canada, Church of England) are very spare indeed. At the other end of the continuum we see churches of the Anabaptist or Free Church expression, which are very exuberant indeed. Between them, a centrist, or moderate, position is held by the churches of Presbyterian-Reformed expression because they are, relatively, more exuberant than, say, Lutherans, and more spare than, say, Mennonites. Methodists would also be moderate in this way, but because they emerge later, they will be discussed later. So, *spare* and *exuberant* do help us in this functional (nonoffensive) analysis as long as we remember that there is a bridge group between, sharing some of the characteristics of both. And, the use of *right, center,* and *left* allows us to avoid using *conservative* and *liberal,* which confuses more than enlightens an attempt to distinguish and define.

Protestant Distinctions

The distinctions between the Protestantisms will come clearer if we discuss the various viewpoints under five headings, some more important than others, but all suggestive of the differences: the Bible, church and state, church architecture, clergy vestments, and the sacraments.

Protestant views on the Bible are most interesting because we start from an apparently identical point for all the Protestantisms. The most essential Protestant slogan is *sola scripture* (the Bible alone). It would seem that all Protestants, from right-wing Anglicanism to left-wing Free Churches, believe the Bible not only to be authoritative but authority enough to challenge the historic church on matters of doctrine. Yet, when it comes down to it, not all Protestants define "the Bible alone" in the same way. Lutherans, Anglicans, and Presbyterian-Reformed churches believe in the authority of the Bible. But when bringing new members into the church, they organize biblical doctrine into agreed upon formulas in creeds and especially catechisms. It is not enough that believers have read, and been inspired by, the Bible; they have to get it "right" according to certain formulas that themselves are historical. So, in this sense, these churches conform to the spare typology because they are disciplined, historical, and corporate. On the other hand, people on the left wing have a direct and immediate appeal to the Bible. When they say "the Bible alone," they mean absolutely alone. There are no other formulas to organize the believer's consciousness, no other authorities between God and humankind. In this sense they are exuberant in that they are individualistic and contemporary ("What does this verse say to you?") in their usage. Moreover, for them the Bible is *exhaustive* in its prescriptions, whereas the right-wing Protestants often hold the Bible is *suggestive* but does not exhaust the possibilities of God's revelation, some of which came through the historic creeds and confessions of the church. Alas, there is no way to resolve this dispute because the Bible itself does not tell a reader whether or not it should be read suggestively or exhaustively.

The issue of church and state also illustrates the distinctions between the Protestantisms. Churches on the right are often referred to as "the magisterial Reformation" because the church was brought in and/or supported by the actions of the state ("the magistrates"). In the Lutheran provinces of Germany, the Scandinavian nations, and in England, the church and the state are seen as equally the creation of God, and, importantly, mutually supportive of each other. In a formulation associated with Lutheranism, the intentions of God are said to be a sword that has two edges, the church and the state. They work together when wielded by God's authority, i.e., those who are in authority are God's chosen ones to rule. On the left wing, the Free Churches try to be precisely that—free from the state. They say that the nation-state and the community of believers are never the same and that their foundations are entirely different. This view is most notable in those of the Free Church expression known as "the peace churches." They are pacific, but their pacifism should be seen not merely as an opposition to war alone but to any and all actions of the state as binding on the community of belief. They would see the Lutheran "two-edged sword" as indistinguishable from Catholic medievalism, and they oppose the notion of a sacral society ("Christendom") in all its parts. For the centrist Protestants, things are not quite so clear. Reformed and Presbyterian churches tend not to see the church and state bound as closely as the Lutherans and Anglicans, but they do see the proper role of secular authority as supportive of sacred purpose. While believing in the institutional separation of church and state, they nevertheless seek to promote the impact of religion on society.

When we move to the artistic and cultural styles of the Protestantisms—how they build their churches and even how they dress their clergy—the differences are quite remarkable. Architecturally, Lutheran and Anglican churches are typically indistinguishable from Catholic churches (indeed, many of them were once Catholic churches). But, either in building new churches or in preserving old ones, they were loathe to change much, if anything. In a Catholic church, when one enters and looks to the front, the visual focus point is the altar at which the central act of worship (the eucharist) is performed. And, horror of horrors to

Free Churches, the pulpit is often on one side with a lectern on the other, where the Bible stands. A clergyperson cannot preach from the Bible, it is said, if he or she does not even stand near it. As to the clergy themselves, the right-wing churches attire them in ways indistinguishable from Catholic priests. For right-wing churches, the Bible doesn't speak on that, and when the Bible is silent, one follows the lead of tradition in the church.

Church buildings of Free Church expression seem to model their ideology of the church. Indeed, many people will not call their buildings "churches" ("the church" is people, they say) but, variously, "meetinghouses," "assembly halls," etc. And, the main room for meeting is not a "sanctuary" (only people are sanctified, not rooms, they say) but an "auditorium." That word tells everything, for in an auditorium there are auditors, i.e., those who hear the preached word ("Word"). Thus, the visual focal point is not an altar but a podium. On the podium is a Bible, to make the point that worship is about hearing the sermon. There will be a communion table somewhere near the front, though it might not be used very often. As to dress, the clergy do not wear what (to them) is antiquated medieval dress, but they follow the fashion of the day, in moderate renderings of what is in style. If they are set apart from ordinary people at all, it will not be in forms of address ("Pastor Bob," not "Father Jones" or "Reverend Jones") or of dress (suit, not gown) but in what they call "the unction" of their preaching.

Reformed and Presbyterian churches take a moderate midpoint between the two positions above. Their churches will typically have a pulpit central in the "sanctuary" (not "auditorium") so that the preaching of "the Word" will be central. But the table (not an altar) will also be in a central position, suggestive of the importance given to the sacrament of Holy Communion. The form of dress is different from both left and right. Not merely the fashion of the day as in Free Churches (although many Reformed clergy became more Free Church in style over the years, nor the medieval garb of Lutheran and Anglican clergy, but a black, academic gown sets them off from ordinary people: They are learned clergy, lecturing, as it were, on the Word of God.

The sacraments reveal the greatest differences between the Protestantisms. Starting from the left wing this time, Free Church Protestants often ask if there is any such thing as sacraments. For most of them, the Bible alone (absolutely and utterly alone) is the guide for faith and life, and they turn to it for guidance on what the historic church called sacraments. The New Testament, of course, does not use the term at all, and (since the Bible is exhaustive and not merely suggestive of God's intentions) if the Bible doesn't say it, we don't do it. They note correctly that the term *sacrament* is of Greco-Roman, pagan origin. In any case, the only sacrifice to be offered to God is ourselves. Now, they do have "ordinances," i.e., ceremonies that Jesus ordained: baptism and the Lord's Supper. But ordinances, while important (baptism *is* important, say, to the Baptists!), are not "sacramental." They are important ceremonies that are symbols of actions that have *already gone on* (the believer already has believed and *then* is baptized; the believer already accepts Christ's substitutionary atonement and *then* is reminded by the ceremony of Christ's sacrifice on their behalf). But, importantly, nothing *goes on* in the ceremony itself. To the contrary, on the right wing, Lutherans and Anglicans believe in sacraments (although not the seven of the medieval church). For them, the Bible is authoritative, but where it is silent or vague, one follows the history of the (inspired) tradition of the church. So, the fact that the term *sacrament* is, by itself, extrabiblical is not a matter of embarrassment. They believe that sacraments are "the means of grace," which, along with preaching "the Word," are the means by which the grace of God comes to humankind. So, while they have symbolic effect in causing the believer to remember the covenant of grace, sacraments point to much more than that. To the horizontal effect, these Christians believe in an added vertical effect. The "action" of belief goes on *both before* (as in Free Churches) and *during* the ministry of the water or the bread and wine. In short, a "miracle" happens in the sacraments because Christ's "real presence" in the bread and wine have an objective reality.

Let us stop the analysis for a moment and go to illustrations to illumine the point of what is meant by the "miracle" of the sacraments. Let me share two personal encounters to which sacramentalist Protestants will say "yes" and ordinance Protestants will say "I have trouble with that." I was asked to be a godparent to the child of a good friend. At the end of the baptism part of the service, the clergyman prayed, "We thank Thee, O Lord, for this daughter of the church who was born again this morning." At the coffee hour afterwards, I took this up with the minister and asked him whether or not any later action on the baby's part would be necessary. He said, "Of course," but, "of course, she will take that action that demonstrates her salvation." But I insisted, "How do you *know* she will?" He asked if I was a Calvinist who believed in God's sovereignty? I said I was and I did. He asked, "Will anything separate this girl from the love of God, will heights or depths or principalities or powers?" I replied, sheepishly, "No." "Then," he said, "what are you worried about? She *has* been born again; she *is* being born again; she *will* be born again." I accept the theological rectitude of that formula, but readers of this book will judge for themselves whether I am right or wrong, even accepting the theology, to worry a bit about the baby's "born again" status. It is not entirely, I think, a matter of me "being of little faith" that I will, as a godparent, feel much better about it all when that girl makes a public profession of faith in her own behalf. But, when (I do not say "if") she does, will the sacrament's miraculous effect have been proven? Once again, readers of this book will have to decide for themselves on that.

In another case, this time on the sacrament of Holy Communion (or "Lord's Supper," or "the breaking of the bread," as other expressions say it), the Protestant readers of Free Church expression may say, "This confirms my worst fears about the crypto-Catholicism of right-wing Protestant churches." All Protestants— right across the spectrum—reject the Catholic theology of "transubstantiation," in which the substances transform to the actual body and blood of Jesus, and that they do so on the formulaic actions of the priest. However, an entire theology hangs in the balance of the prefixes *tran* and *con* because Lutheran and Anglican theologies of the sacraments insist that a special miracle of grace comes *with* ("consubstantiation") the sacraments. And, even though the substances don't transform (in the normal definition), they do carry with them a special and tactile grace that is well beyond the realm of symbol. Yet, Free Church evangelicals react in horror to the suggestion that grace "happens" in this way. In fact, however, Lutheran and Anglican readers of this book will wonder what the problem here is. To them, the idea of a "symbol" is a kind of religious Neoplatonism, in which "reality" is best known through "the idea of reality" rather than the apparent, physical things of reality.

Back to analysis, then, it would seem that views on the sacraments punctuate most clearly what separates the various expressions of Protestantism. To the left wing, the right's views on the sacraments cause the former to wonder if the latter are Protestants, even Christians, at all. To the right wing, the left's Neoplatonism and euphemisms about sacraments cause the former to wonder if the latter can be taken seriously when they insist that they are "Bible-believing Christians." In the functional terms we are using in this chapter, we see most markedly the utility of the words *spare* and *exuberant*. The moderate centrists of the Calvinist-Reformed-Presbyterian expression (and later on the Methodists) are ambivalent about the degree to which a miracle "happens" in the sacraments. They are torn between the teaching of Calvin, which tends toward Luther, yet not *consubstantialist*, and that of Zwingli, which tends toward the Free Church, yet not "ordinance." The Reformed viewpoint emphasized the *action* of grace, not the nature of the substance in the sacrament. Somehow calling it a sacrament without being exactly certain what the substances in a sacrament are, however, will not satisfy those who hold to the clearer positions of right and left.

Let us conclude this discussion of the Protestantisms before going on, in the next chapter, to see the sociopolitical outworking of the Reformation and Counter-Reformation. In this chapter we have seen the Reformation as intimately connected with the Northern Renaissance, and we insisted that it does no violence to the religious/spiritual nature of the Reformation to do so. It is well for us to remember the context of ideas and movements and the way in which there is interplay between "the times and the person." As Mark Noll, Nathan Hatch, and George Marsden wrote in another connection: "While Christianity always transforms culture, culture always transforms the Christians within it." For us to see the Reformation as part of the Renaissance spirit does not praise or blame Protestants but rather sees them as they were. As we engage them and their diverse movement—whether in agreements or disagreements on the important issue discussed above—we do well to practice those Christian virtues of humility and charity.

Chapter 8

THE REFORMATION AND COUNTER-REFORMATION: SOCIOPOLITICAL ACTIONS AND REACTIONS

There were, in fact, two Reformations, not one. As discussed in the previous chapter, the Reformation was a powerful revitalization movement in the Christian church. Not only did it contribute to the renewal of the church in the form of the new denominations and sects of the Protestantisms but also in the renewal and clarification of faith and practice within the historic church. To say "Counter-Reformation" is already to put a Protestant construction on what happened in the Catholic church after 1517. Perhaps it is better to call the latter the "Catholic Reformation." But, in truth, it was a reaction to the actions taken by Protestants that clarified religious life among Catholics. In short, it was the challenge of the Protestants that caused Catholics to redefine what they *did* accept and *did* believe. In this chapter, we will use a few pages to note the matters of concern within the Catholic church itself. The bulk of the chapter will illustrate this pattern of action-reaction in the sociopolitical realities of two nations—Britain and France.

The Catholic Reaction

We have already noted an important point about Protestantism: that it used the language of liberty to rebel against the historic, institutional church. As discussed above, Protestants *disbelieved* and *disobeyed* in order that they might *believe* and *obey* rightly. But begging the question of authority does not really answer it. The Lutherans and the Calvinists did not intend to bring in a new dispensation of pluralistic toleration. Rather, in their places and times they were as authoritarian as the authority structures against which they rebelled in the first instance. While Protestants had raised the whole question of "right" authority, it was that question, more than anything else, that animated the clarification of belief and practice in the Catholic Reformation.

Later on, and in the context of the American Revolution, one Massachusetts loyalist remarked that he thought it a poor exchange to rebel against one tyrant, three thousand miles away, only to replace him with three thousand tyrants, one mile away. This attitude can be no better illustrated in the Reformation's context than in the career of George Witzel (1501–73). He had originally become a follower of his fellow German, Martin Luther, in hopes of reforming the church. But, he was disenchanted with Luther and especially with "Lutheranism." He believed there were sound grounds for saying that Luther led a rebellion against the pope in Rome only to see himself installed, in fact, as the "pope" of Germany. Witzel reunited with the Catholic church by the late 1530s and became a priest. But even committed to remain within the church, he nevertheless continued to push for reforms *within* the church like those the Protestants were articulating *outside* the church. Witzel's creative suggestions for reconciliation fell on deaf ears on both sides of the divide. Correcting abuses was, by the 1550s, widely agreed to among Catholic leaders, especially in northern Europe. But, the problems for Witzel and other reconciling reformers always turned on the essential problem of authority. Witzel, in the tradition of the Christian humanism of Erasmus, believed in broad toleration of differing views within the whole of the church. But the sixteenth and seventeenth centuries were not ones characterized by a desire for pluralism and toleration among either Protestants or Catholics, as readers of this book who know about the vicious persecutions of the Anabaptists will readily agree.

Many Catholic leaders, especially in northern Europe, called for a general council of churches to deal with the Protestant challenge, in hopes that reforms might be effected—and the departing brethren might be brought back—before battle lines were hardened and it would be too late. The papacy, however, dominated as it had been by southern Europeans, especially Italians and Spaniards, resisted the calling of a general council and only summoned one under some duress. It met during the period 1545–63 in the north Italian city of Trent, in three sessions interrupted by plague, wars, and the deaths of three popes. If one recalls the pattern of representation in the Council of Trent, the outcome might well be predicted. It was dominated by southern Europeans from the beginning, and while their majority decreased over the various sessions, the number of persons allowed to vote on substantial matters never fell below the three-quarters domination by Italians. In fact, when the council first met, twenty-one of the twenty-three bishops were Italian.

Although at first the council was conciliatory, positions hardened on both sides, and, in the end, there were no concessions given to the Protestants. Witzel's suggestions for reform and toleration were rejected. The work of the Council of Trent is very important because it "fixed" Catholic faith and practice until the next council, not called until 1869 (Vatican I). On the basic Protestant slogan, *sola scripture,* Trent replied that Scripture and tradition were one. On the other Protestant slogan, *sola fides* (or, justification by faith alone), Trent replied that while faith might begin the process of salvation, it could be completed only by the sacraments, rightly and duly administered by the church. On practical matters like venerating relics, retaining Latin as the only official language of worship, and the right of the church to grant indulgences if it chose (although the sale of indulgences *did* decline), the Catholic church did not appear to give an inch. And, importantly for Protestants, the Council of Trent even more firmly entrenched the power of the Italian-dominated papacy. In short, the Council of Trent froze the Catholic church—in its doctrine and its practice—at a particular point in time, and most attempts at a thaw were resisted until the 1960s when Pope John XXIII "opened the windows" at Vatican II.

By the 1560s, therefore, the battle lines had been hardened. The Protestants, for their part, viewed the decisions of Trent as confirmation of their leaving a church that was apparently not open to any substantive change. The long history of structural (not moral) reform within the Catholic church was stopped by the freezing of doctrine and practice at Trent. Moreover, the confidence and militancy of Catholics themselves increased after 1563, as witness the growth of a new, militant order, the Society of Jesus (the Jesuits), founded by a Spaniard, Ignatius of Loyola in 1540, just before the Council of Trent began to meet. Under the Jesuits, a worldwide system of missions, education, and training was begun, and the Jesuits can be credited with faithfully spreading the positions of the Council of Trent throughout the international church, with Jesuit priests often the watchdogs of orthodoxy.

Working Out the Meanings of the Reformations

While doctrinal and institutional matters *are* important, it is also important for us to recall that it was in terms of social and political concerns that we see the meaning of the working out of the Reformations. In the previous chapter we have already mentioned the role of sociopolitical thinking, especially nationalism in the emergence of Protestantism. The other side of the same coin is also important to recall: The development of Protestant churches as "religious" bodies had the effect of stimulating social and political thinking. And since, in the argument of this book, it is the method of humanist/Protestant rejection of "authority" we have found interesting, it is difficult to see how the two matters can be separated. In short, once a person or group began to act "like a Protestant" in one area of life, it was easy to see how they would behave that way in other parts as well. If Protestants were prepared to take religious authority "into their own hands," it is likely that political authority would soon follow and *vice versa*. On the other side, once Catholic persons and groups clarified what it meant to act "like a Catholic," they would soon see the implications for other areas as well. Since, in the terms of this book, Catholics submitted to "God's authority" in church and society, they would work hard to defeat those persons and groups whose actions opposed the structures of the world that God had made.

Now, as we try to seek the *meaning* of the Reformation and Counter-Reformation, we see that a focus on doctrine and on churches is inadequate. What is at stake is no less than the determination of who is in charge in church and society, "God" or "Man." How one defines "God" and "Man," of course, is the difficult point since both sides insisted they were on "God's" side and their opponents on "Man's." The following makes the distinction too starkly simple, but it begins the conversation we need to undertake here. Catholic (and sometimes Lutheran): "This world is God's world. He is monarch of the universe. Those in authority carry God's authority, and mere mortals obey those in authority as though they obey God." Protestant: "This world is God's world. He is monarch of the universe. He has given certain principles in his Word that guide the affairs of nations and people. When we, and kings, obey those principles, we obey God. Moreover, it is the responsibility of kings and people to bring the institutions of church and society into conformity with the principle of God's Word." These viewpoints, one quickly sees, are not easily reconcilable. Put far too simply, but in one sentence, Catholics (and some Lutherans) saw the institutions of church and society as "what is, *is* what ought to be"; Protestants saw those institutions in terms of "what is, *is not* what ought to be but what *might become* if essential principles were adhered to." Or, put another way, for Catholics, God's intentions are made known to humankind through institutions that have developed over time; for Protestants, God's intentions are made known to humankind through the Bible, whose most essential point is "the covenant." Institutions, therefore, are God's institutions if, *and only if,* they abide by the principles of the covenant.

The covenant is an organizing idea for most Protestants, both in church and in society. In religious terms, God reveals himself to a people and says, "I will be your God." Moreover, God obliges himself to certain principles by which his "Godness" and their "peopleness" will function. He lays down terms of reference that are as unshakable as God himself; and the people in reply are to fulfill their obligations to each other and to God. Then, and only then, can a true community exist, when the people obey "the rule of law" as faithfully as God does. In societal terms, the meaning of this covenantal ideal was slow in emerging. But in the 250-year period after the time of Luther and Calvin, Protestants in northern Europe (and emigrants to North America) would see its societal emergence tied to its more formally religious definition.

It will be the Calvinists of Reformed and Presbyterian persuasion who give impetus to social-covenantal thinking more than other Protestant groups. When we recall the Protestant continuum, it is possible to see why this would be so. The right wing of the Protestant Reformation was sometimes called the "magisterial Reformation" because the church, in this view, is part of a sacral society in which the "princes" have the obligation to institute and maintain "right religion." This is a type of "Protestant medievalism" in which religion and society are aspects of a coherent whole. Luther's image is helpful here, that God's intentions are like a sword that he wields. It has two edges, the church and the state, and both move together to accomplish God's will. It is almost inconceivable, in this view, that there would be a conflict between church and state because that would imply a conflict between the twin purposes of God. Thus, while Lutherans and Anglicans *did* think about societal issues, that thinking typically emphasized bringing the church and the state into harmony.

On the other end of the Protestant continuum, the Anabaptist and Free Church expressions founded their very identity on the separation of the church and society. While some were "quietist" (accepting social matters as given because "real" reality was spiritual or otherworldly), others were "activists" (especially the "peace churches" who courageously confronted the power of the principalities). But the common belief among all Anabaptist–Free Church people and groups was that the community of belief was *never* coincident with the state or any other secular definition. Since their societal thinking emphasized the separation between religious and political institutions, their valuable contribution was of little positive value in seeking that very conjunction of institutions they sought to separate.

The middle ground of the Protestant continuum is a relatively more ambiguous position than either "magisterial" or "Free Church." The Calvinists would neither dissociate religion and society nor say they were the same; the church was to be neither dependent on the state for origin and support nor separated from it in the doing of God's "kingdom service." In spite of—indeed, possibly because of—the ambiguous and ambivalent position, there was a great deal of creative political thinking among Calvinists as to just what the relationship of religion and society is and ought to be.

John Calvin (1509–64) not only lent his name to a movement within Protestantism, but the experiences of his life bear markedly on the questions we have raised. When we later get to the discussion of Britain and France, we will see the influence of Calvin's Geneva. So it is right for us to take a few glimpses at the career of this influential reformer. French by birth, he was attracted to the scholarly life through the classics, and his early intentions were to be a humanist Catholic scholar, much in the manner of Erasmus of Rotterdam. But, while a student at the University of Paris in the 1520s, some Christian-humanist friends introduced him to the works of Martin Luther. Although he was not interested in Protestantism for nearly another decade, the association with certain Christian-humanist scholars would eventually lead him along a trail of ever-increasing public exposure. He preferred the life of scholarship, which seemed to suit his retiring personality. In his various flights to safety, from France to Switzerland and back several times, he gained a reputation and following in Basel, Strasbourg, and, above all, in Geneva.

The city council of Geneva had, early in the 1530s, decided to wrest authority from the city's Catholic bishop, whom they expelled. At the urging of William Farel, Calvin came to Geneva. But the magistrates were no more interested in the Farel-Calvin religious imperatives for city life than they had been of the bishops', and Calvin was expelled in 1538. After Calvin lived again in Basel and in Strasbourg for a while, the city council—afraid of a radical swing toward a more "democratic" Anabaptism or a counterswing back to a vengeful Catholicism—invited Calvin to return, either as a moderate among more radical options or the least of three evils. From 1541 until his death in 1564 Calvin was an increasingly influential citizen of Geneva. But, it must be emphasized, he never possessed direct political power, both because he thought it wrong to seek it and because the city council would not allow it. Calvin never advocated *theocracy,* the rule of the state by the church. He did write certain "ordinances" and "constitutions" by which he thought the city government ought to run society in conformity to God's Word, but he did so as a private citizen who believed that the church and its pastors ought to be the conscience of the state.

The "rule of law" in the Geneva of Calvin's time was rigorous, and probably caused the city to have a higher moral tone than other cities its size. Sometimes the regulation of moral life by the city council was minute and even ridiculous (against gambling and laziness, and even to forcing a man to make a public confession for naming his dog "Calvin"). But the definition of "moral" was broad and deep, involved in what we would call social justice issues. For example, the regulation of wages, fair interest rates, providing medical care, and even developing a sewer system were very important to the welfare of citizens, as much as the regulation of doctrine. God's Word, Calvinists insisted, spoke to all areas of life. As this form of Protestantism spread, it became known as revolutionary because the covenant given by God to humankind does not admit to distinctions of public and private, sacred and profane, religious and secular. In this view every knee, both personal and institutional, would one day bow in declaration of Christ's lordship. So every task was a "gospel" task, in that every task worked toward the coming kingdom.

While all Protestants believed in the covenant as an organizing principle, the Calvinists pressed it further than other groups of Protestants. Indeed, it is in the realm of the social translation of the covenant that Calvinists showed the full potential of Protestantism, and, in respect to that, many Catholic leaders saw what had to be done in terms of reaction. There is an intimate connection between the idea of religious covenant and the corresponding idea of society governed by contract. It is not possible for us to think of a developing constitutionalism—the rule of law—except in the context of religionists who also saw "law" as governing all aspects of life. We can see this if we take the example of Britain, in which the ideal of government by social contract was victorious, and of France, where it was decisively defeated.

The British Case

Protestantism's arrival in Britain is an embarrassing story for the Protestant cause. The monarch Henry VIII was worried, above all else, about the precarious hold his family had on the throne. He wanted a male heir, in the belief that his daughter would never be able to hold on to the power his family had so recently won. Had he known what formidable women his daughters, Mary and Elizabeth, would be, he need not have worried. He was, moreover, along with his friend and advisor Thomas More, an early and outspoken opponent of Luther. But he led the church in England out of the Catholic church for other than theological grounds by various actions in the 1530s. So the church in England became the "Church of England" under the monarch's headship at just about the same time as Calvin's rise to prominence in Geneva. Many English church leaders studied events in Geneva and tried to define the new church of England in Reformed terms (Anglicanism's defining document, "The Thirty-Nine Articles of Religion"). Without going into the various swings of the story between 1547 and 1558 (the death of Henry, the short "Protestant" reign of Edward VI, and the abortive and violent "Catholic" reign of Mary), Elizabeth I (1558–1603) came to the throne determined to bring religious peace. The Church of England became, by early seventeenth century, a broad church, in terms of tolerating various views within the church and a middle way between Catholicism and continental Protestantism.

It was in the seventeenth century, however, that the real story of our theme of conjoining religious and political change was to be played out fully. Both in the Parliament and in the church, there were elements that wanted to limit the prerogative of the king and that wanted to move the Church of England in a more determinedly "Protestant" direction. While they were not always the same persons, they drew on a common impulse: wanting to "purify" both political and religious institutions. These "Puritans" soon came to see that the two causes were very much intertwined. The cause of parliamentary ascendancy picked up momentum in the reign of James I and came into open combat during the reign of Charles I (1625–49).

In the "Petition of Right" (June 7, 1628), the British Parliament asserted its authority in a way in which no national legislature had previously done. It insisted on the rule of law, obliging both monarch and people to an agreed upon social contract. Charles tried to evade this parliamentary assertion of rights, and he was in great difficulty after the Reformed church in Scotland refused to give up its Calvinistic theology and structure in response to a royal decree to have the Church of England become uniform in England, Ireland, Scotland, and Wales. He wanted the army to put down the rebellious Scots, but the Puritan-Parliament forces reminded him of the obligations to live within the law, namely that Parliament controlled all appropriations. Charles decided to let all this go no further, and he assembled a loyalist force near Nottingham and awaited combat with a parliamentary army centered in the southeast of England and with a Scots-Presbyterian army. In the English Civil War the two aspects of covenantalism were joined on one side and the two aspects of authority joined on the other. The Puritan-Parliament faction, under Oliver Cromwell, was to be successful even to the point of executing the king. Regicide—the killing of a king—proved too radical a measure for most Britons. After a short period of a Calvinist "republic" (what else would heirs of the Greco-Roman tradition establish?) under Cromwell, a military group seeking stability encouraged the Parliament to restore the monarchy, and the Stuarts were brought back from their exile in France in 1660. From that date to their flight in 1689, the two rulers of the house of Stuart presided over something of a replay of the events of the first half of the century. The two Stuart monarchs, Charles II (1660–85) and James II (1685–88), soon came to see what their friend Louis XIV in France had come to see—that a truly "Christian" king had to oppose the "covenantalists" in both forms, Protestant and Parliament—and they bent their efforts to restore authority in both matters. Louis XIV, for his part, pledged secret support to Charles II in this effort, in the hope of restoring Catholicism and true political authority in Britain.

The dynastic situation of royal succession becomes complicated in the 1680s, but it is enough to say that Parliament refused to accept the rule of James II when his second wife gave birth to a son, who was baptized a Catholic. James had hoped then that the people would accept the fact that, in due course, they would once again have a Catholic king. All that had to happen was for the Parliament to do nothing. But Parliament turned to the Protestant daughter of James II (by his first wife, also a Protestant), Mary, who had married William, of the house of Orange, in the Netherlands. William and Mary were invited by members of Parliament to lead an invasion force into Britain as a result of the flight of

James to France in search of French aid to quell the Parliament. When William was crowned monarch along with Mary, he was required by Parliament to say a phrase (which all British monarchs have since said at their coronation) that brings the religious and political issues together finally and openly. He pledged to uphold "the Protestant and Reformed faith" and "the liberties of Parliament." It was, apparently, only in a fully Protestant, even "Reformed," nation that the rights of the rule of law could be preserved.

William was not allowed to linger long in London because the exiled king, James II, was to mount a military campaign with French aid. James knew that the struggle to restore his kind of "Christian" monarchy (Catholic in religion and without a Parliament) would be long and hard. So he chose to invade through that part of Britain that had remained steadfastly Catholic—Ireland. With a popular base of support, he thought, his armies could quickly conquer Ireland, and from there they could move on to the other parts of Britain. It was a sound strategy. Had he been able to control Ireland, it would have given his plan a good chance of success. Had he been successful in the re-Catholicizing of Britain, he could have joined his colleague Louis XIV of France in Louis's grand scheme of rolling back the Protestant tide. Louis believed, and probably rightly, that Britain and France working together could have attacked the Protestant provinces of Germany and the Netherlands, then gone on to the Scandinavian (Lutheran) nations.

It was with all this on the line that William of Orange—only recently king of Britain—marched at the head of an army across England and Wales, sailed across the Irish Sea, and marched to the extreme northwest of Ireland to meet the forces under James outside Londonderry in 1690. William's Protestant army was successful at Londonderry and the decisive victory over James's forces came later in 1690 along the River Boyne in north-central Ireland. Indeed, the Battle of the Boyne is one of those major turnings in religious history (small wonder that the Protestants of

Northern Ireland today have a sense of their importance in religious history). It ranks along with Tours, where, in 732, Christian forces under Charles Martel had stopped the advance of Islam into Europe. It is possible that Catholicism might have been restored universally and Protestantism marginalized had events turned out differently in central Ireland in the spring and summer of 1690. In the end, of course, the dream of Louis XIV to restore "proper" authority to Europe was not to be realized. Deprived of his English ally, he could not advance against other Protestant areas. The constitutional monarchy of Protestant Britain is one of the great legacies of the Reformation. The story of Protestantism in Britain, begun so ignobly about 150 years earlier with Henry's dynastic difficulties, turned out to be a noble triumph for Protestantism and a bitter defeat for Catholicism.

The French Case

Across the English channel, in France, a story similar in character was to have a different result. John Calvin had great hopes for Protestantism in his native country. As noted earlier, he had worked in Strasbourg for a time, and even though he was living in Switzerland, his hopes for France were high, as witness his dedication of *The Institutes* to the French king, Francis I. In the unstable conditions of mid-sixteenth-century France, religion and politics were quickly mixed, and the decision of a person to become a Protestant was as much a political as a religious act. The vast majority of Protestants in France were Calvinists, called Huguenots; and, although many of them were placed in high or key positions in society (about half the nobility and a quarter of the cities became Protestant), their numbers were small in the total population, never rising above 10 percent.

True to Calvinist ideology, Huguenots allied themselves with members of the nobility who sought to limit the power of the monarchy. But their minority status caused them to adopt a stance—a hoped-for toleration of their religious existence—that was to be the key to understanding the story. In England, the discussion of religion and politics was dominated by a woman, Elizabeth I (1558–1603). In France, the key person was also a woman, Catherine, who after the death of her husband in 1559 was the real power behind the throne successively occupied by her three sons (Francis II, Charles IX, and Henry III) between 1559 and 1589. Always intent on playing off the various factions of ultra-Catholics and ultra-Protestants, she offered toleration to the Huguenots in hopes of bringing stability. But fearing the rise of Protestant power under Conde and Coligny, and especially prince Henry of Navarre, she secretly worked with the always powerful Guise family to strike a decisive blow against the Protestants. On August 22, 1572—Saint Bartholomew's Day—Catholic forces coordinated attacks on Huguenot leadership, and some 20,000 Huguenots were killed, 3,000 in Paris alone.

It is in response to that massacre that Huguenot-Calvinist theorists began to develop a theory of revolutionary activity. Calvin himself had been relatively conservative politically and had roundly condemned the "conspiracy of Amboise" in 1560, when some Huguenots had tried to kidnap King Francis II. Yet, later Calvinist theorists drew from Calvin's writings a theory of resistance to tyrannical authority in which lower officials could be called upon—in the name of covenantal, higher authority—to resist wrong authority and restore proper authority. It was important to them that such resistance not be merely a popular uprising but a restorationist action led by members of the duly constituted government. Pamphlets by Francois Hotman (1573), Theodore Beza (1574), and Philippe Mornay (1579) encouraged a pattern of political thinking that suggested that a social contract existed between the rulers and the ruled. But when the high rulers disobeyed their part in the contract, lesser rulers should move to restore the proper terms of the contract, in the interests of all the people.

Affairs outside France directly impinged on the situation in France. Spain, a bastion of Catholic conservatism, had been defeated by Protestant Britain in a major naval battle in 1588. King Henry III of France saw this as the time to restore stability to France by working with the Huguenots to reduce the power of the Catholic League (an anti-Protestant faction controlled by the Guise family) in Paris, because he saw the league as part of a Spanish attempt to control France. Sensing the possibilities that the defeat of Spain's armada might offer, Henry made common cause with Prince Henry of Navarre, the most prominent Huguenot and brother-in-law of the king. King Henry was assassinated in 1589, resulting in the ascension to the throne of Henry of Navarre, now King Henry IV.

Here is one of the great turnings in religious and political history. Henry IV, like Elizabeth I across the channel at the same moment, was a *politique*. That is, he wanted to end religious and political strife, and social stability was more important to him than trying to make France uniformly Protestant. In any case, he reasoned, there was little possibility of doing so. Therefore, the middle way was attempted: He realized that France could only be united under a Catholic king, but a king who respected the place of Protestantism in France. In 1593, he publicly renounced his Protestantism, and the nation rallied behind him, ultimately vastly reducing the power of the Catholic League and Spanish influence. The Huguenots had a stake in the new order, too, in the form of the Duke de Sully, a prominent Huguenot, who became Henry IV's first minister of government. Most notably, Henry IV issued the Edict of Nantes in 1598, which made religious tolerance (although only for certain denominations) the official policy of France. This was a radical policy, as toleration was not officially guaranteed anywhere else in the Christian world at the time.

Henry IV was assassinated in 1610, and the situation in France came unglued again. The two most important persons in seventeenth-century France were not kings but cardinals in the Catholic church. Because Louis XIII and Louis XIV were both minors when their respective fathers died, the queen mothers appointed two cardinals as regents, Richelieu (1624–43) and Mazarin

(1643–61). They were both clever and able to do some unexpected things. Rather than moving quickly against the Protestants, as one might have expected, they kept the policy of Nantes (toleration) alive in return for the cooperation of the nobles and the Huguenots in consolidating French authority at home against Spain and Austria. Yet, even so, they slowly eroded Protestant liberties at the same time as they eroded the local liberties of the nobility. In 1649–52, there was a series of uprisings (known as the Fronde) led in the countryside by the nobility. Perhaps encouraged by the success of the British parliament in the English Civil War at the same time, Huguenots joined with the nobility in this uprising against royal authority. Successful at first (Mazarin had to flee France for a while), the nobility could not govern France well (their ideology, after all, was localist, not nationalist). With the coming to power of Louis XIV in 1661, the pendulum swung back toward centralization of power and toward Catholic predominance.

Louis XIV need not be discussed much here because the main texts treat him fully. But, for our purposes, let us mention that he believed deeply that the religious and the political questions were really only one question—the issue was authority, in both the church and state. It was difficult, if not impossible, for him to conceive of a monarchy that shared power with a parliament (Estates General) or of a church in which true authority was not centrally controlled and universal. Thus, it comes as no surprise that Louis saw his historic role as that of restorer of authority, in all its aspects. For him, 1685 was to be the key year: he revoked the Edict of Nantes, and he pledged support to his friend James II of England. In both cases the goal was the same: to extirpate Protestantism and to restore proper political authority. Many Huguenots preferred to renounce Protestantism rather than lose their places in French society. Some tried to keep Protestantism going as an underground movement. Many fled to Protestant areas where they would be welcomed or where toleration was practiced (the Netherlands, Prussia, Switzerland, South Africa, Britain, Northern Ireland, and Philadelphia, Pennsylvania, where the Quakers guaranteed toleration). After 1685, while not disappearing completely, Protestantism in France was spent as both a political and a religious force. For Louis XIV the examples of the Fronde in his own country and of the civil war in England convinced him that if a king was to be the sort of king God intended, then both religious and political authority had to be centralized.

The Challenge of Jacques Bossuet

In concluding this chapter, let us make reference to two pamphlets that appeared nearly at the same time in London and Paris: John Locke's *Of Civil Government* (1690) clearly reflects the theories of constitutionalism that had emerged victorious in Protestant Britain; Jacques Bossuet's *Politics Drawn From the Very Words of Holy Scripture* (1688) clearly reflects the theories of "divine right of kings" that had emerged victorious in Catholic France. Locke's views are, by far, the better known because they are among the foundational views of liberal, democratic thought and practice. The nature of society, says Locke, is a contract between all persons to regulate their common affairs. Principally, the contractual parties are the king and the people, who mutually oblige themselves to abide by the terms of the contract as stated in law. When either party exceeds contractual bounds, law assesses penalties, and, if the monarch refuses to be bound by the contract, it is the right and the responsibility of the people to oppose the monarch in order to restore the covenant. In a later chapter, we will return to Locke. But it is enough for now to say that his work is of the first importance as a major statement in the intellectual history of the West on behalf of government by social contract. Yet at the same time, Locke's work is a reflection of the basic tension of the Protestant movement on the vital question of authority in religion and in society. To assert that both church and state are proper churches and states when they live by the terms of the covenant-contract is all very well. But, the question remains as to how the terms of the contract shall be known and interpreted. In short, another rendition of the humanist-Protestant dilemma: Where does authority really and truly reside?

Jacques Bossuet had a different way of looking at the whole question. In the terms developed in this book—though not spoken in this form by Bossuet himself—the basic problem with all forms of covenantalism was that they were disguised forms of humanist assertiveness. For him (again in our terms), the problem lay in mere humans trying to take to themselves the authority that rightly belongs only to God. In short, is "man" the measure of all things, or is "God"? Since most of the readers of this book are Protestants, and since all accept liberal, democratic theories over "the divine right of kings," it is important to work through Bossuet's views.

The Protestants, be it recalled, chanted their most essential slogan ("the Bible alone") into Catholic ears. So it is with great cunning that Bossuet gives his title, *Politics Drawn From the Very Words of Holy Scripture.* It was as though Bossuet would challenge the Protestants on their own ground, saying, "All right then, you believe in the Bible as the only authoritative guide for faith and life; let's see what the Bible says about politics."

The Bible, of course, is not a primer for political, economic, or social activities. It is, first and foremost, a testimonial to the nature of God and a revelation of what God wants done in the world he created, i.e., that all people might name his Name and follow him. Along the way, to be sure, there are glimpses of what God wants in terms of political, economic, and social arrangements. But the main point of the Bible is to declare the glory of God and to show humankind in proper (deferential and dependent) relation to that glory. Bossuet pointed out that, both in overarching example and in proof text, the Old Testament has several types of arrangements—kings and charismatic leaders—through which God rules the people. But, that *is* the point, that whoever is in charge is in *God's charge* and answerable only to God, not to the people, at least in any substantial sense. Moreover, the Old Testament actors who defied the rule of God's anointed ones are portrayed as the deviant participants in the story.

As we move to the New Testament, the story is similar. Jesus' kingship was not earthly, and even Jesus recognized the legitimate rule of the secular state. Furthermore, Paul, in writing to Christians in Rome (people living under adverse circumstances) enjoined them to obey authority. One looks in vain for footnotes at the bottom of the page in the Epistle to the Romans where Paul might have stated exceptions to the norm of obedience. Rather, one sees that God is still, and always, in charge of human affairs and that humans must obey those who God has (apparently) put in charge. Even if a regime might appear "evil" (first century Christians could relate to that), what humans might count for evil, God might count for good. Humans, in their temporal finitude, cannot see the long run; God, and God alone, in infinity, can see the long run and will accomplish things according to his purposes. In the meanwhile, God knows who is in temporal authority and (apparently) "allows" it: to believe otherwise is to doubt the authority of God. So, Jacques Bossuet concludes that to question the king is to question God; to try to impose a "legal" restraint upon God's appointed one is to do it to God; to try to divide sovereignty between king and people in a parliament, under a "rule of law" is to elevate human creatures to the realm of creators. For Bossuet, there *was* the possibility of a "bad" king, i.e., one that does not please God. But, that was for God to deal with, not the people. In any case, in the Christian West, a true king *would* defend right religion in the very acts that also defended right authority. In doing so, the church and the state would work together to disclose God's intentions in sacred and secular concerns. God, in the end, is the king of creation and monarch of the universe. That divine monarchy has delegated, as it were, divine right to the two monarchs of this world, the pope of the church and the king of a nation-state.

For Jacques Bossuet, for Louis XIV, and for Catholic Christians, the revocation of the Edict of Nantes in 1685 and the determination to roll back the tide of Protestantism in northern Europe was the logical conclusion of the above train of thought. How could we tolerate, they asked, the existence of a religion that denied God's authority in all its parts? How can we countenance the existence of Protestant nations when we believe that they will spread their disease of human arrogance to other (Christian) nations? For them, what the Protestants called a Counter-Reformation was the real Reformation. It was their beliefs and actions that really and truly represented God's truth and what God wanted in his world. For them, the Protestants were not really Christians at all but people who would use the Bible for purposes of humanist assertion. For them, the taking of God's prerogatives into human hands was to open the way for others to go further, until, in ensuing generations, everyone would do that which was right in their eyes and there would be no agreed-to moral compass for society. Such a "liberal" state may satisfy humans temporarily, but it would displease, even anger, God eternally.

The two reformations might not have taken the course they eventually took had a willingness to find common ground existed in the early sixteenth century. A Catholic church council in 1512 had called for reforms in the church's spiritual life. Also, the early sessions of the Council of Trent did offer certain compromises (marriage for the clergy, the Bible in common languages, end of indulgences). But, Protestant leaders rejected such overtures as too little, too late. In the end, the positions of Trent hardened as had those of the Protestants.

In conclusion, then, the reformations must be seen as developing. The views discussed here developed over time. They were not solely theological (in the narrow definition) but were about *the meaning* of theology in social and political contexts.

Chapter 9

THE SCIENTIFIC REVOLUTION: RATIONALISM AND THE MODERN WORLDVIEW

We live today in the modern world—of that there can be no doubt. While there are many foundations of modernity, the most important is the set of understandings about ourselves and our world that we call scientific. This chapter will deal with the scientific revolution and with the development of rationalism.

In our time there has been an overusage of the word *revolution* (industrial, political, social, even sexual revolutions are said to have happened). But if *revolution* means substantive change, then the scientific revolution of the sixteenth and seventeenth centuries can be said to be *the* revolution of the modern world. Whereas once the main ways of speaking about life were philosophical-theological, in the modern world the main metaphors are scientific. The prestige of scientists in our time is like that of theologians and artists in a former time. Nor is it easy to say where it happened because the scientific spirit is not confined to one nation. It is precisely an international community of "disinterested" researchers that did—and do—the main work we call scientific. This revolution is, in the Western world, pervasive in its impact.

Secular, Scientific Humanism

This direction of thought leads us to one of the most important points in this book. We are trying to think (as self-consciously as possible) "Christianly" about Western civilization, and early in our thinking we made something of a distinction between Hebrew-Christian and Greco-Roman ways of thinking and believing. In the humanism of the Renaissance, as we have noted, some later Christians want to see the opposite of Christian thought. We cautioned against seeing Renaissance humanism as the antithesis of Christianity. While *that* humanism had the potential for a secular end, we insisted that it was not a *necessary* end. To have insisted on that, we said, was to have written the Protestant movement out of the story. Yet, it should be recalled that the potential for secularity was always present in humanism, if not fully realized in the Renaissance-Reformation period. In the time of the scientific revolution, however, we can see that full potential for secularity being realized. In order for humanism to be *secular humanism* it must be *secular-scientific-humanism*. And, it is important to say that this sort of humanism *does* present a real challenge to Christian belief, *not* because science itself directly supplants religion, but because a worldview known as *rationalism* arose from science. It is on that worldview that all modern systems of political, social, and economic institutions are based.

Over the next pages—indeed, over the next few chapters—the reader will notice that we talk little about the church and its activities. We must, in intellectual honesty, follow the main line of the Western story, and the church has not been the main line in the past three hundred years. We must allow this story to be told, not because we believe it to be the "right" story, but because it is the most far-reaching story in describing the modern world. So let it be said once here, and not at each chapter heading, that as Christians we do not think the modern worldview is normatively correct. The assertion of "man" to take the center place from "God" is not a world picture we can endorse. The modern philosophy of secular-scientific-humanism has come to grief in our time, as Christians thought it might. But, in honesty, we cannot stand on the sidelines of this story and carp, "I told you so," because that would be a dishonest and uninteresting exercise. In the final chapter, we will ask if Christians can accept the modern world. But in asking that we will by then know what we have to give up if we give up on modernity. A few sentences ago I said that *all* modern political, social, and economic systems are based on secular-scientific-humanist worldviews, and I do mean *all* (in the West). So, what we have to give up is our own as well as someone else's pet schemes and arrangements: capitalism no less than communism, conservatism no less than liberalism, our nationalist egos no less than theirs. It is in respect to facing up to modernity that we will see what is involved in that facing up and what being "kingdom seekers" may actually cost us. But we are running ahead of the story here, so we should return to the task at hand while, at the same time, realizing the importance of it.

Science and Its Revolution

Science, as is well known, is not a philosophy of life but a method by which the natural world can be studied, described, and known. Internal to that method are certain procedures and understandings that it is well for us to discuss. *Empiricism* is so basic a part of scientific inquiry as to leave moderns saying, "Well, of course." By empiricism we mean that way of looking at life that says reality can be known—and *only* known—through observation and by experiment. Every time someone *does* science, that scientific work is predicated on a prior acceptance of empirical reality as "real" reality, and no scientist may leap to a conclusion that is beyond what can be known empirically. Moreover, logic must govern the reasoning steps in the process, so that the steps of increasing knowledge are integrally connected. But not any sort of logic will do because science depends on *induction* not *deduction*. With *inductive* logic internal to the scientific method, one reasons from what is known to the next stage of what can be known empirically and then on to a generalization carefully inferred from the evidence. *Deductive* logic is as integrally connected as inductive (it *is* logic, after all). But deduction's starting point is a general category, and the reasoning process is from general to particular.

As we complete our definitions we must discuss another term, not *necessarily* connected to the scientific method but *said* to be connected; the term is *rationalism*. Now, contrary to science (a method), rationalism is a worldview, a cosmology. As Crane Brinton states well, rationalism is "a cluster of ideas that add up to the belief that the universe works the way a man's mind works when he thinks logically and objectively, that therefore man can ultimately understand everything in his experience as he understands, for instance, a simple arithmetical or mechanical problem. The same wits that showed him how to make, use and keep in repair any household contrivance will ultimately, the rationalist hopes, show him all about everything." This is a *cosmology* in that it offers an explanation about life in a way that science, as science, does not (indeed, cannot). While rationalism is clearly based on science, it is not a *necessary* conclusion to draw from science. The perennial questions in which humanistic scholars are interested (truth, goodness, beauty, justice, God, or "god") are not answerable by science, but they are very much the terrain of rationalism. In the rest of this chapter we will discuss science in the sixteenth and seventeenth centuries and then try to see how the rationalist cosmology evolved.

While it is unimportant to ask who began the scientific revolution, it is nevertheless typically said that Nicholas Copernicus (1473–1543) was the first modern scientist. His major work, *On the Revolution of Heavenly Bodies* (1543), was, as historian Thomas Kuhn suggested, a book not so much revolutionary as one that made a revolution. In saying this we mean that Copernicus's research in astronomy began a way of thinking about the place of humankind in the universe that was in marked contrast to the prior theories that had governed thinking for most of the Christian era (e.g., Ptolemy, who worked around A.D. 150). The difference can be simply stated, but the impact is enormous: Whereas Ptolemy—and all Christians for some 1,400 years—insisted that the earth was the center of the universe, Copernicus's work suggested that the sun was central and that other "heavenly bodies" revolved around it. The importance of Copernicus's work is that mathematics was the governing discipline for the empirical organization of other data. Galileo Galilei (1564–1642) was not a mathematician but a natural philosopher with great mathematical skills. He was the first astronomer who actually used a telescope to investigate the heavens. Not only did Galileo's work support the Copernican hypothesis, it demonstrated a far greater complexity to life in the universe than had previously been imagined. Moreover, to Galileo it proved that life in all its complexities was a matter of mathematical regularity. The book of life, he said, is written in mathematical symbols and in order to understand the story in the book we need to understand those symbols.

The method of mathematics was developed further by René Descartes (1596–1650). While Descartes is famous for a number of works, his contribution to the scientific revolution lies in his insistence that the only credible intellectual authority is human reason. Even though, as a deductionist, he worked against the grain of modern inductive science, his insistence on the primacy of reason in *The Discourse on Method* (1637) was a major way station in the development of the modern mind. Descartes's contemporary, Francis Bacon, was not really a scientist at all but a student of language and learning. His contribution to modernity lies also in the realm of method. He often likened himself to Columbus, helping to lead the way to discovery of a world more complex than previously known. Firmly an inductionist, he emphasized the role of empirical knowledge in experiment and innovation. For Bacon (in *The Advancement of Learning,* 1605), "idols" prevented humankind from advancing. While there were several kinds of idols, an important one was the "idol of the tribe," in which people thought that most truth had already been discovered. To those who said that God had revealed himself, Bacon insisted that that was a limiting notion because it blinded people to the changing and the new. Bacon believed that a new method of knowing would liberate people from the tyranny of the already-known and free them to develop new capacities and capabilities that would advance the frontiers of knowledge so that nature might be controlled by humankind.

Isaac Newton (1647–1727) was perhaps the most famous person of the scientific revolution, and his *Mathematical Principles of Natural Philosophy* (1687) stands as the magisterial work of the new science. Deriving his ideas from the work of Galileo and Kepler, Newton noted that the force of gravity was relational, i.e., that gravity on earth related to the gravitational pull between the planets of the solar system. He demonstrated that relationship mathematically, and as his theories of universal gravitation became well known they seemed to seal a conviction that had emerged in the scientific revolution: that the natural world was one of regularity, consistency, and "law." Moreover, the more one was open to the possibilities of discovery, thought Newton, the more one would be impressed with the regularity and the consistency of nature. In short, the more one knew about the nature of things, the more one *could* know.

These persons are the founders of modern science. While others could be mentioned, these give us a fairly clear picture of the nature of the new worldview. Science, we have said, does not in itself provide a cosmology (an explanation of things). But we can see how this new attitude toward and from science encouraged people to think in terms of a new general explanation. It is important to recall that all the scientists mentioned in this chapter were Christians, and they believed that their emergent, new "world picture" did not diminish belief in a purposing and powerful God. As Newton insisted, scientists deal honestly in cause-and-effect relationships between observable phenomena; but when they push their findings back to first causes or forward to final causes they will find the Jehovah God in which Jews and Christians have always believed. The world may be seen in mathematical and mechanical terms, but origins and conclusions must be seen in spiritual terms. Yet—and this is a most difficult point—the God who begins and ends is seen, in the new world picture, as remote from the processes that operate between first and last causes.

The image of a great clock is often thought helpful in describing what we mean here. God, it is said, created the universe. But since God is infinite, he created it with infinite complexity. At the same time, that complexity conforms to a regularity, to which God obliges himself as much as he obliges humankind. The whole universe, from the very large to the very small, is like a great, cosmic clock, the likes of which no one had ever seen nor could even imagine. As humans began to see the complex, interconnected, and patterned working of that clock, their initial reactions were awe and excitement at the possibility of being able to learn about it. This initial group of scientists felt what all of us felt when we first lay in a field and beheld the sky on a clear night and then later looked more closely at it through a telescope in an observatory. They felt what all of us have felt when we first looked into a microscope and saw what we could not see with our eyes alone. The world is wonderfully and awefully made, and human response is wonder and awe. Yet the rules (natural laws) by which the clock operates can be learned about and lived by but cannot be changed. Our wishes and our wills are no match for natural law. We can sit under an apple tree in the autumn, but we cannot will the ripened fruit to fall up when it leaves the branch. It will always fall down. We can drive down a highway at high speed in our cars, but we cannot will our bodies to stay put when we crash into another object. We may well go through the windshield. Just as gravity is a "law" of nature, so are there other "laws": "bodies at rest tend to stay at rest, bodies in motion tend to stay in motion"; therefore, if my body is traveling in my car at a certain speed, it will continue at that speed even if the car stops abruptly—hence, my exit through the windshield.

Science and a Personal God

Christians, for their part, have always believed in a God who creates and ends time. But that is not nearly enough to say about God, at least for most Christians. Most believe in a God who intervenes in a personal way. However satisfying first and last causes may be theologically, they leave us alone in the universe in the intervening period between first and last. In this context, prayer becomes irrelevant; most Christians are very uneasy about that. Should we ask God to reverse or even abrogate natural laws because we ask? Should the laws of thermodynamics not apply to a situation about which we feel deeply? Should the natural life cycle of gestation, birth, growth, maturity, decay, and death not apply to us because we cannot bear the loss of a loved one? Specifically, shall I not be thrown through the windshield of my crashing car if I call on the name of the Lord just before impact? In a rationalist's view of the world it would be almost insulting to God to say, "I know you have created the glorious and complex world according to your purposes, but would you change its pattern of operation for me if I asked you?" In Newton's view of the world, one could almost hear a divine reply: "I am sorry that you, or your loved one, has to die in this way, and it breaks my heart to see yours breaking. But if I changed the rules for you, where will that leave everyone else, except in an unstable, even capricious, universe in which no one would know where they stand?" If Christians can begin to see how God cannot violate the rules he created for the world, how much more would non-Christians find irrelevant Christian assertions about a personal God who intervenes in the great and mundane affairs of our lives?

Most working scientists today have no interest in the reconciliation of religion and science because, as empiricists, religion to them is not so much wrong as irrelevant. But scientists who are Christians—from Galileo to Newton to scientists in our time—feel a deep burden to reconcile the two. Yet, their way of reconciling is essentially what Newton himself said three hundred years ago: that God obliges himself to natural law as much as we must. God created the world according to laws he reveals to us. We cannot change any more than we can challenge those laws. In any case, the world *is* orderly, precise, and coherent, and it does no good to try to ignore or to subvert what God has clearly put in place. To try to do so is, in effect, to tell God that he cannot run his universe according to the plan that he apparently conceived before the beginning of (what we call) time. However, such an apparently unassailable conclusion has left many Christians deeply troubled about the nature of the world and about their place in it. They are troubled because, of necessity, the God of first and last causes cannot be summoned to intervene in their lives. Their prayers cannot be intercedings on behalf of a special plea but only praisings in recognition of a God who purposes all, according to a general plan not to be changed by special arrangement. This was in the seventeenth century, and even today is, a matter of deep concern to many Christians, and the conclusion to the debates surrounding these concerns is not in sight.

The whole concern of Christianity's relation to science is difficult and complex. But, as we already noted, science—as science—does not ask, nor is even interested in, questions of human values and purposes. In fact, the supposed conflict between science and religion really occurs when science is transmuted into rationalism, which is not a method, like science, but a worldview, or cosmology. When Carl Sagan, a prominent scientist in our time, writes that "the cosmos is all that is or ever was or ever will be," he is engaging in precisely the transmutation we are describing, of taking science and making it into a worldview. He, and many others before him, take the method of knowing we call scientific and insist that all life is like that. Moreover, rationalists insist that nothing of value can be known except by the empirical and inductive methods. In thus trying to explain the world, they deal not only in present realities but, implicitly, in first causes and last things.

A Worldview Shift

In the next chapter we will examine this worldview in its programatic and prescriptive aspects—the Enlightenment. But for now it is enough to assess where we have come. We now see that it was too much to ask that the medieval Christian worldview disappear immediately in the Renaissance. There was still too much

Christendom around for that to happen so quickly. When we look for the roots of modernity, we must see a process of transition that went on in three stages from roughly 1350 to 1750. At base, that transition is the formative one in the quarrel between ancients and moderns, or between a world in which truth is revealed and a world in which truth is discovered. In the former world, humankind may lose sight of the truth or even go against it. But, nonetheless, the truth *was* the truth, and since no one who knows truth could be right to be against it, it mattered very much that truth be defended. In the latter world, various truths could be—indeed *should* be—tolerated because there was no such thing as a whole, single, revealed truth. Humanity was discovering truth, and it was exciting to come to different realities in nature and society. Why, the moderns asked in this latter world, should we hold people to outmoded ideas? Why not let them stand on each others' shoulders and reach the higher for all humankind?

The first of the transitional stages was Renaissance humanism. Its application of a methodology of questioning authority was the beginning of a corrosive process that resulted, in the end, in the crumbling of Christendom. It questioned authority very self-consciously, and artists and scholars were able to do new things because of the liberation they felt from prescribed behavior and belief. Some of them rebelled against medieval authority in the name of the free individual and many of them in the name of replicating the glories of antiquity. Its importance was that it opened the mythical Pandora's Box.

The second of the transitional stages from medieval traditionalism to modernity—the second out of the box—was Protestantism. It, too, was a corrosive to the solid structures of prescribed thought and practice. But of greater importance to our present concern, Protestantism was an agent of change because it allowed people to question foundational religious beliefs. Moreover, Protestantism split up into several large, mostly nationalist, churches and into smaller groups and sects. As such, it encouraged skepticism because, if one could legitimately challenge universal truths and institutions, one might conclude that there was no such thing *as* universal truth.

The third stage from tradition to modernity in thought—Rationalism—was not merely corrosive but destructive of medieval Christendom thinking. Perhaps the destruction of medievalism was only possible when humanism and Protestantism had done their corrosive work first. Not only did Rationalism marginalize the supernatural realm, it placed humankind in a material and mechanical universe. Whereas Christians responded to revealed truth (Protestants had no quarrel with this, it was just the sources of truth that were in question), the rationalists found truth in patient investigations, encouraged and emboldened by the mathematical certainties of their discoveries. But, like the box that Pandora opened, many vices came out as the result of her curiosity. And, along with the human vices came hope. It was hope for this world that marked off the modern from the traditionalist. And it was—and is—secular hope for humankind that is the hallmark of modern thought. This Enlightenment, to which we turn in the next chapter, was a worldview as different as imaginable from that of, say, St. Augustine, St. Thomas Aquinas, St. Francis of Assisi, Martin Luther, John Calvin, and Jacques Bossuet.

Chapter 10

THE ENLIGHTENMENT:
A WORLDVIEW IN ACTION

The term *Enlightenment,* like *Renaissance,* sounds positive. From the time of revolution in science to our own, the Enlightenment has suggested a positive development: a new source of "light" had been discovered by humankind and that light would allow people to behave differently than had previous humans. As historian E. J. Dijksterhuis (1961) has suggested, the essence of the Enlightenment was the belief that the world could now be seen in mechanical terms and defined in mathematical language. Moreover, this new picture of the world could be seen by intellectuals and ordinary people alike. As we mentioned in the previous chapter, if one mastered the method of knowing how to operate and to keep in repair the merest mechanical contrivance, that same method—on a grander scale—would allow humankind to know and to explain the microscopic and macroscopic workings of the universe. It was not merely that scientific phenomena were like that, but *all* reality was like that, i.e., subject to immutable and implacable natural laws. In this chapter we will examine the workings of this world picture, and its principle spokesperson, on the way to a conclusion that will wonder if *Enlightenment* is a positive term after all.

Rationalism is, as we have noted, a worldview based on science but not necessarily the logical conclusion of the scientific method. In the Enlightenment we see the full development of rationalism as worldview. Let it be clearly understood that the Rationalism of the Enlightenment is based on an analogy, more asserted than demonstrated. The analogy insists that the world of plants, animals, and things (nature) works a certain way, i.e., according to "natural law." And therefore, Rationalism argues, all of reality is like that because all reality similarly operates by natural laws, if we had eyes to see them. The world of human affairs, in sum, is the same as the natural world because the same laws that govern each govern all. The task for Enlightenment thinkers, therefore, was to ascertain those general laws that governed all reality and then apply them to the various cases that came up, whether political, economic, social, or religious. The key persons in making this "mathematization" of the world clear to their generation and ours are Isaac Newton and John Locke. In gauging their importance, Crane Brinton offers the following, stunning sentence: "Together, Newton and Locke set up those great cluster of ideas, Nature and Reason, which were to the Enlightenment what such clusters as grace, salvation and predestination were to traditional Christianity."

From Science to Society

In the previous chapter we discussed Newton at some length, so there is no need to repeat the importance of his work again, other than to restate for emphasis that *The Mathematical Principles of Natural Philosophy* is one of the most important works in creating the modern worldview. Two chapters ago, we spoke of John Locke's *Of Civil Government* as being of primary importance in the history of the development of liberal, democratic constitutionalism. Newton had a decisive influence on Locke. When Newton's *Mathematical Principles* appeared in 1687, Locke was said to have been simply astounded by its implications. But, a cautious man by nature, he wanted to be sure of the mathematical side of Newton's work. Before going any further, he asked his friend, Dutch mathematician Christian Huygens (1629–95), to verify through replication Newton's equations. When Huygens agreed with the accuracy of Newton's mathematics, Locke knew that he had a degree of scientific certitude on which to proceed in his own thinking about personality and about society. Taken together, their work had revolutionary potential, and the impact of Newton and Locke on Western thinking has been very great indeed.

The generation of social thinkers that followed Newton and Locke made a leap—from science to society—that the two themselves were not ready to make. Writers of books, pamphlets, and, generally, makers and molders of opinions, moved beyond Newton and Locke quite quickly to assert certain things about persons and society. "Nature," they said, is the external world in which we live, but all things in that world are not "natural." Indeed, many things that exist are actually "unnatural" because they do not conform to the natural laws of the universe. As the work of Newton was filtered to ordinary readers, it became expressed as follows: The world is orderly and coherent, operating by natural laws. The more we know about those laws and regulate our lives accordingly, the more the world will be a better place and we humans will be happier. Moreover, as people are increasingly liberated from unnatural ideas and behaviors, they will see what they had not seen before, that this new way of thinking and behaving is no more, or less, than common sense. "Reason," in this common definition, would enable humankind to find "natural" institutions, i.e., personal, social, economic, religious ways of behaving that will place us all in harmony with each other in the way in which plants, animals, and things are at harmony in the "balance of nature."

If the seventeenth century had witnessed the revolution in science, the eighteenth saw the implications of "natural Rationalism" worked out for society. The Enlightenment is the literary output of persons in several nations—although France was the most important—who confronted political and religious authorities with the social translation of the ideas we have, above, associated with Newton and Locke. These writers and advocates corresponded with each other and commented on each other's works. They had a sense of solidarity in a cause, i.e., of liberating the human spirit. They are often referred to as the *philosophes*.

The Program of the *Philosophes*

Chief among the *philosophes* was Voltaire (his real name was Francois Marie Arouet, 1694–1778). His work was so offensive to the political and religious establishment in France that he was jailed for a time, and he did most of his writing outside France, earlier in London, later in more tolerant Geneva, where French authorities could not touch him. Like many of the continental *philosophes* he had great admiration for the British system of constitutional monarchy and Britain's toleration of different viewpoints in intellectual discourse. Voltaire's most important work was *Elements of the Philosophy of Newton* (1738), in which he, more than any other writer, popularized the salient points of *The Mathematical Principles*. He wrote in nearly every *genre,* producing prose, poetry, history, and plays. His most remembered work was *Candide* (1759), a satire on conventional political, social, and religious thinking. By appropriating Newton's insights for the social task, Voltaire is a vitally important bridge figure in bringing the abstract and logical work of the scientific revolution to bear upon the relationships of people in society.

While Voltaire was the most important single Enlightenment thinker, there were many others. But the Enlightenment translation of science to society is best seen in a collective effort called *The Encyclopedia,* a twenty-eight-volume work, the first of which appeared in 1751, the last in 1772. If there is a compendium of Enlightenment thought, this is it. More than one hundred writers contributed to it, on varied subjects from attacks on "unnatural" religious and political institutions and behaviors to practical articles on canal building and scientific agriculture, showing once again that the new way of thinking was both intellectually respectable and practical. *The Encyclopedia* was deeply resented by religious and political leaders who tried, unsuccessfully, to prevent its publication or to censor it. Well did traditional authorities, especially in France, fear *The Encyclopedia* because it set forth the most advanced, even radical, social, political, and religious ideas heard or read in Europe up until that time. The work of Diderot and d'Alembert, the editors of *The Encyclopedia,* is one of the enduring monuments of the Enlightenment.

While we may cite various writers as typifying the Enlightenment, it would be misleading to see it in an ahistorical way, i.e., as a single body of thought. In fact, Enlightenment thinking developed over time, and we can discern at least two, possibly three, generations of main representatives of Enlightened thought. What we can see, almost, is a replay of the distinction between spare and exuberant humanism, with an earlier generation of Enlightenment spokespersons being cautious and a later generation claiming rather more extravagant possibilities for the new ideas. Whereas Newton retained a deep Christian faith, David Hume questioned the faith in all its aspects, including faith in reason; whereas Locke's case for a constitutional arrangement for society was moderate and historical, i.e., retaining but reshaping older institutions, Rousseau was a thorough-going radical who believed that justice would only come in a fully democratized society. It is important for us to keep this distinction in mind. In assessing the Enlightenment in its prescriptive, programmatic, later phase, we will take the examples of political and religious thinking (leaving economic thinking for later discussion in the context of the industrial revolution).

In political thinking we can see both development among the *philosophes* and disagreement as to what natural institutions should replace the unnatural ones they observed. It was all very well to say that monarchy and nobility were not natural bases for proper politics. But, the question was—and is—what sorts of arrangements should be put in their places? We turn to France for insights into Enlightenment political thinking because the obviously unnatural institutions in place there gave rise to more political writing than in other nations and because, with the writers of *The Encyclopedia* centered in Paris, there was a critical mass of political thinking.

The work of Baron de Montesquieu (1689–1755) is widely regarded as a link between the relatively conservative thinking of Hobbes and Locke in the seventeenth century and the relatively radical thinking of Rousseau, later in the eighteenth century. *The Spirit of the Laws* (1748) was both analytical and prescriptive. In true Enlightenment fashion, Montesquieu believed that law undergirds all reality. And, while the laws governing human behavior were complex and more difficult to discern than physical laws, they were no less operative. Observing that there were really only three forms of government—monarchic, republic, and despotic— he suggested that whatever history a particular nation had would lead it in one of those directions. For France, however, he left analysis and became prescriptive. As an admirer of the British system of "crown and parliament" (he had lived in Britain for several years), he hoped France might adopt for itself what he believed he saw in Britain. (Even though we now know he saw somewhat incorrectly, his prescriptive ideas still have great impact.) He believed that political liberty could only be safeguarded if there was a definite division of powers between the executive, legislative, and judicial branches of government. In the French case he saw a special role for the *parlements* (a kind of aristocratic court system) in restraining the power of the monarchy. While clearly no democratic radical (the aristocracy had an important function in his idea of a constitutional monarchy), his work was to have great impact on a revolutionary generation that, fifty years later, was to try to write constitutions, most notably in Philadelphia in 1787 and in Paris after 1789.

While Montesquieu's work is interesting and influential, it is not intellectually controversial. But it is around the work of Jean-Jacques Rousseau (1712–78) that controversy has swirled for the past two centuries. By the time of Rousseau, i.e., the later generation of Enlightened thought, it was no longer necessary to state the natural law foundation of society because that was no longer a novelty. Rousseau was a radical (Latin, *radix,* "the root") in that he tried to get to the root of the matter of modern society. If a society was good, he said, there would be good people, because the purpose of society was, in his view, to provide a context in which people might become better. He probed deeper than any of his contemporaries into the question of what the "good life" is, or ought to be, once people were liberated from unnatural institutions. In Enlightenment language, he saw the good life for persons as already related to the goodness of society and, importantly, that the good life for individuals could not be realized unless and until their personal destinies were bound to the destiny of society as a whole.

In *The Social Contract* (1762) Rousseau offered not only a foundational document for modern thought but, more specifically, the classic statement of modern *democratic* thought. Until Rousseau, most political thinkers in the Enlightenment mode thought of society as an aggregate of individuals. Rousseau, however, looked at this proposition from the other side. While he agreed with earlier Enlightenment thinking, to the extent of his ridicule of unnatural institutions (especially historically sanctioned inequalities), he saw individual persons as having worth in *relation* to each other in society. So *The Social Contract* begins with the ringing phrase, "All men are born free, but everywhere they are in chains," and he insisted that unnatural chains be broken (the metaphor of breaking chains was to be picked up by Marx eighty years later). But the breaking of chains was not an end in itself because that would result merely in the freeing of atomistic individuals.

Rather than an *atomistic* theory of society, as argued by much of Enlightenment political thought, Rousseau developed an *organic* theory of society (drawing on Plato, and even Calvin) in which persons *became* worthy persons when they acted together in the society of (what the Romans had called) "virtue." Democracy, therefore, could be the only rational form of government because only through equal participation in society could people develop the full potential of their humanness. The best thing for society and persons within it could be known through what he called "the general will." This is not mere majority rule, because the majority might still believe and behave unnaturally. Indeed, the general will may be known only by a few persons at first. And those with such knowledge have a responsibility to share the ideas of virtue—and, controversially—*force* change in society to conform with the general will. We will return to this problem—which is the central problem of modern democracy—in two other connections (the French Revolution and the Russian Revolution). But it is enough for now to note that when Rousseau says that freedom is obeying the general will (even if one has to be forced to be free to obey) he is saying the same point, in a secular context, as Augustine and Calvin did in religious contexts: that perfect freedom obtains when a person follows the will of God. In short, a larger will (or "Will") outside oneself, whether from society or from God, is the stuff of which personal liberty is, paradoxically, made. At once we can see what is controversial about Rousseau: Deterministic systems, whether developed in the religious context of Geneva (Calvin) or in the secular context of London (Marx), applauded Rousseau's insights into the need for a societal definition of virtue. Liberal systems, whether developed in Virginia (Jefferson) or Glasgow (Adam Smith), feared that Rousseau's insistence on the general will might result in a society as despotic as the unnatural one they opposed in the first place. In short, at stake here is whether or not Rousseau is the father of modern, democratic, liberal thought or of modern, totalitarian thought. For example, if I voted for a person *not* elected president, I *must* believe (in a modern democracy) that the general will ("virtue") had been expressed by the majority in the person elected. If I did not believe that, I would have to foment revolution after the election. It would seem that the stability of any democracy turns on an acceptance that the general will is right. Listen to Rousseau:

In order, then, that the social compact may not be but a vain formula, it must contain, though unexpressed, the single undertaking which can alone give force to the whole, namely, that whoever shall refuse to obey the general will must be constrained by the whole body of his fellow citizens to do so: which is no more than to say that it may be necessary to compel a man to be free.

We will return to Rousseau again because no realistic analysis of modern society can, or has, gone on outside of the concerns he raised. Whether or not he is the hero or the villain of modern political and social thought can be argued. But, in any case, his concerns cannot be avoided.

As we move now to religious thinking in the Enlightenment, we see a similar development over time, i.e., from the more moderate essays on toleration by Locke in the 1690s through the development of deism to the challenges to supernatural belief in Voltaire, Hume, and Lessing. As the world became more "enlightened" (i.e., scientific-rationalist), one of two things, it was said, would occur for religion; it would accommodate the new mode of thinking or be outmoded by it.

Accommodation usually came first in the guise of toleration. We should remember that early in the eighteenth century the norm for Catholics and almost all Protestants in Europe and in its North American colonies was one of intolerance and conformity. Christianity's multiplicity of denominations and sects was unacceptable to most religiopolitical establishments, with a few notable exceptions (for example, briefly in Catholic Maryland, Baptist Rhode Island, Quaker Pennsylvania, where Christian co-religionists of other denominations were allowed a relatively free exercise of religion).

As the eighteenth century matured, however, it became apparent to those who had drunk at enlightened waters that if there was a religious "message" from rationalism, it was, at least, that all truth is universal, so that denominations and sects were philosophically unacceptable. This is so because, if the various groups insist on the monopoly of right thinking and even have social and political penalties in place for those who do not conform, they are acting irrationally and unnaturally. Surely, the argument went, if

God, who is One, created the world according to his master plan, it does not stand to reason that he would reveal that plan more to Anglicans than Presbyterians and Methodists (England), more to Lutherans than Mennonites (Germany), more to Calvinists than Baptists (Massachusetts). The program of the religiously enlightened complemented their political program. On the continent of Europe and in Britain and British North America, political and religious authority often went hand in hand whether in the power that clergy had or, conversely, in the exclusion from power of "dissenting" groups. So, the plan for religious toleration in Locke's writings, for example, emanates from the same spirit as his political writings.

Deism is the name given to rationalist religion. Given the rationalist's view of nature and of humankind, their formulation of religion is unsurprising. They believed that God could be known through his work in nature and that belief would issue in moral behavior because the more one was in harmony with natural law, the more one acted morally toward one's neighbors. Religion was "disenchanted" in that the supernatural and mysterious were rejected in favor of the natural and the explicable. The God of nature or, as Thomas Jefferson put it, "Nature's god," was the God ("god") of all, whether they be Christians of any denomination or adherents of other world religions. The unique in various religions was attacked by the *philosophes* (for example, David Hume's attacks on Christian miracles as empirically unverifiable). What they sought was a tolerant, morally useful religion that would unite, not separate, all humankind.

This direction of thought leads us to the conclusion of this chapter. In the initial paragraph of this chapter we wondered if, upon analysis, we would be able to see *Enlightenment* as a positive term. We are not now going to abandon our pledge to avoid praising or blaming, but there are a few comments that need to be made by way of summing up. It seems inappropriate, given the argument of these past two chapters, to pit science against religion. As we have seen, science is not, and cannot, provide a worldview. Only in the attempt to make science do what it cannot do (Rationalism) can a cosmology be provided. Moreover, Rationalism makes certain assumptions about God and about humankind in the universe, that, taken together, are a kind of religion. So we come now to the first part of the conclusion: We who are Christians must not allow the setting of a false agenda that states that religion is outmoded and scientific-rationalism is progressive. We insist that Rationalism be called for what it is: an alternative religion. It is not merely playing with words to insist on this, but it is following the logic of Rationalism itself. It is a "religious faith" in that it gives explanations for the origin, development, and future of humankind. It might hope to start with empirically verifiable facts, but it goes well beyond the realm of facts to cosmic explanations about life.

Rationalism: A Counter-Religion

Rationalism is as much a "faith" as the Christian faith it tried to supplant. This is no more apparent than in the Enlightenment's most precious doctrine of progress, as much a "doctrine" as any Christian doctrine, i.e., based on faith assumptions. In Condorcet's *Progress of the Human Spirit* (1794) there are stated, importantly, ten stages through which humankind has progressed to the then present, when humans had within their power to attain a state of perfection. It is on the doctrine of progress that Christians and rationalists part company most profoundly in religious terms. Christians believe that humans have lapsed into an irremediable condition that can only find remedy outside the reach of human action and outside the reach of human time. Rationalists believe that whatever lapses there have been in human behavior were not inherent in humanity but environmentally engendered. Therefore, the remedy *is* at hand and within the scope of human action and time. As historian Carl Becker suggested in his *Heavenly City of the Eighteenth Century Philosophers,* what the rationalists did was to take the traditionally Christian expectation of heaven and secularize it, bringing it into "this world" (Remember Sagan's dictum, that *this* cosmos is all there ever was or will be?). They appropriate Christian expectations by using familiar notions about a hoped-for future perfection. And they believed that with the increasing knowledge about nature and humankind, it would be possible to "progress" to that highest of all stages of development.

Christians, of course, will not be persuaded by this religion in which the two cities of which St. Augustine spoke—the city of God *(civitas Dei)* and the city of this world *(civitas terrena)*—are melded into one. Now comes the second part of our conclusion. It is important for Christians to realize, in a way that many Christians since the eighteenth century have not realized, that they must not be on the defensive on behalf of religion against irreligion. The Enlightenment faith *is* a religion—a counter-religion to be sure, but a religion nonetheless. Moreover, Christianity *did* continue to flourish in the period since 1750. The mere mention of the great successes of Methodism is enough to substantiate that. But, the difficult point for Christians in our time to recognize is that traditional Christianity, both Protestant and Catholic, while continuing and flourishing, became increasingly marginal in a world whose institutions were based on the religious faith of the Enlightenment. As mentioned earlier, *all* forms of modern social, political, and economic institutions in the West are developments of the Enlightenment ideology. While traditional Christianity continued to exist after, say, 1750 and even enjoyed great success in North America while experiencing some declines in Europe, it was to become an increasingly minority faith. The main contours of developments in Western civilization were increasingly founded on the alternate faith presented by the Enlightenment.

Chapter 11

THE AGE OF DEMOCRATIC REVOLUTIONS: THE NORTH ATLANTIC WORLD "TURNED UPSIDE DOWN"

When, in 1781, Lord Cornwallis surrendered to the victorious Americans and their French allies at the battle of Yorktown, he ordered his military band to play the tune "The World Turned Upside Down." The American Revolution did, in some ways, turn the world of the late eighteenth century upside down. But more than the American Revolution, the French Revolution was to turn the world in different directions. Moreover, there were attempted revolutions—mostly thwarted—in Belgium, Holland, Geneva (Helvetic Republic), Ireland, and Poland. These various movements, taken together, represent more than isolated and unrelated episodes. In fact, we can see a larger pattern throughout the North Atlantic world of a desire to put into practice the ideas and attitudes that sprang from the Enlightenment world picture. We see a desire to move from theory to practice, from the concerns of the few to those of the many, or, as historian R. R. Palmer suggests (1972) to bring "the Enlightenment into practical politics." In this chapter we will look into the nature of revolutionary ideas and, after a brief inquiry into the American and French revolutions, assess the meaning of the revolutions. The essential questions we want to ask here are, were these revolutions really a practical Enlightenment, and did they have a lasting impact in turning the world "upside down"?

Rationalism and Revolution

The ideas we have been following in preceding chapters—the worldview of rationalism—moved from being the possession of a few (the *philosophes*) to the possession of the many. This was accomplished by the popularization of the rationalist ideal through books, pamphlets, and sermons. There were broad movements throughout the North Atlantic culture to bring discussion of political, social, and economic issues into the open. While it would be fruitless for us to try to untangle the dilemma of whether or not ideas cause revolutions or revolutions generate ideas, we can observe that ideas gain currency and immediacy in times of social stress and upheaval. While it is technically accurate for Alexis de Tocqueville to have observed about the French Revolution that "vast though it was, it was entirely unforeseen," when the revolution *did* happen the ideas supporting it surprised no one for they had been discussed for several generations, especially in France and Britain. It would be wrong for us to insist, in a narrow definition, that the Enlightenment "caused" the American and French revolutions. But at the same time, and in a broad definition, Enlightenment impulses provided the basis for revolutionary thinking, in that it caused people to believe that the world could change for the better as all persons began to live according to natural law.

If the American and French revolutions were truly Enlightenment revolutions, what principles would they have advocated, and what actions would they have accomplished in order to qualify? As developed in the previous chapter, the program of the Enlightenment was its politicization. An Enlightenment revolution, as such, would accomplish the goals of the program, i.e., to break the structures the "old order" *(ancìen régime)*, which was characterized by monarchy, social stratification, and established churches. Since crown, class, and church were seen as the essential structures of established order, to bring in the new order of natural and rational arrangements one would have to undo those structures for liberal and democratic (free and equal) institutions to emerge. Both the American and French revolutions, therefore, seem to qualify (at least for now) as Enlightenment revolutions, because the net result of both was to usher in a new world in which a return to the old order of "establishment" was impossible. The French Revolution is more complex and has had greater continuing impact than the American, but we take the simpler American Revolution first in the hope that it will offer us an analytical perspective with which to examine *the* revolution of the modern world, the French Revolution.

The American Case

The American Revolution is, in fact, curiously named, because if it indeed was a revolution it was not characteristic of those society-wrenching activities we associate with other modern revolutions, among them, the French, the Russian, and the Chinese. There were no reigns of terror, no mass deportations, no forced labor camps, no major social reconstructions, no dividing of great estates among the peasantry. What we today would call "a war of national liberation" seems to have taken place. While it is sure that a new nation-state came into being after 1776, it is doubtful that most of the "revolutionaries" themselves thought of themselves as making a revolution.

So, on one level of thinking, it is possible to think that an American "revolution" never happened. On another level of thinking it is possible to follow Carl Becker's idea that there were, in fact, *two* American revolutions: The first American Revolution, he suggested, was about "home rule," while the second was about "who should rule at home." While most historians no longer think Becker to be correct about the two social revolutions he thought happened, his direction of thought remains helpful for our analysis of the events in British North America between 1770 and 1791.

We can identify two markedly different groups of American founders. While they made common cause in rebelling against the

British Empire, the respective reasons for their rebelling were different, and for each, the "meaning" of "their" revolution was different from that of the other group. One group adhered to the main tenets of the Enlightenment, while the other found the more moderate political renderings of the Renaissance more appropriate. The one believed that the whole tendency of history was that humankind would progress in the direction of a liberal democracy; the other disliked arbitrary power in monarchy but thought that a "republic of virtue" would be a safeguard against the king and the mob alike. We must choose our terms of reference carefully here: *liberal* and *conservative* won't do because one group was functionally conservative (not changing things much), while being ideologically liberal (believing in democracy), whereas the other group was functionally liberal (changing the structures of things) while being ideologically conservative (disbelieving in democracy and preferring the rule of persons with "a stake in society"). Nor can we use *democrat* and *republican* because the group believing in direct democracy called themselves republicans throughout the revolutionary period, while those believing in a republic called themselves federalists. Instead, let us use the functional and inoffensive terms—ones more appropriate to the late eighteenth century—the *popular government group* and the *strong government group.*

We discuss the popular government group first because its importance is first chronologically, though the chronology is not one of neat sequence. The popular faction in the revolutionary movement believed that "their" revolution was about bringing in a democracy, or the rule of all by all. The emergent leader of this group was Thomas Jefferson. At the Continental Congress in Philadelphia in 1776, he was selected to draft the Declaration of Independence. When he removed himself from the assembly to write the document, it is noteworthy that he took with him a copy of *Second Treatise of Government* by John Locke, whose phrase "life, liberty, and property" found its way into the Declaration as "life, liberty, and the pursuit of happiness." When Jefferson produced his draft Declaration, some observers said it contained "nothing new." Jefferson replied that he had not intended to write anything new but to couch the American claim to independence in terms of reference that everyone knew. His argument was rooted in Enlightenment ideas. Truths that are "self-evident" are precisely that and need no special pleading to defend them. Nature (or, as he said later, "Nature's god") had endowed all persons with rights that could not be "alienated" from them. Neither government nor society had given persons their rights; rather, rights were inherent in their human condition. Moreover, government must protect those rights. Citizens need not obey a government that, in extreme cases, abuses those rights. The purpose of government, in short, is to provide a context in which individuals may pursue "happiness" as they see fit. This is as complete a statement of Enlightenment goals as one is likely to find.

The spirit of the Declaration—its Enlightenment ideas—was put into practical structures in the first "constitution" of the United States, the Articles of Confederation. The Articles, while not formally ratified by all the states until 1781, were the effective constitution of the United States from 1777 to 1789. Given its Enlightenment basis, the structure of government under the Articles is predictable. Importantly, it was a *confederacy*, a term that would come up again in American history. For this group of patriots, "their" revolution was about opposition to national, central government (Britain) and about a determined fragmentation of governmental power among its parts. The reason for this fragmentation was principled: It was thought that a democracy was possible only when the levers of power were close to the hands of the people themselves. While there was a "national" government under the Articles—it did, after all, defeat the British Empire—that government saw effective power residing in local and state governments. Moreover, the leaders of the popular government faction distrusted the whole notion of elitism and thought it inappropriate to "their" revolution—a democracy.

The strong government group was no less patriotic than their popular government co-revolutionaries. While there may have been a large number of loyalists to Britain who eventually, and after the revolution, found their way into this group, the main group was faithful to the patriotic cause. For them, however, "their" revolution was not about creating a localistic democracy but about creating a "republic of virtue" in which there would be liberty, to be sure, but, as John Adams said, an "ordered liberty" under the rule of law. While this group shared the popular group's fear of "tyranny" in London, they also feared what they called the rule of the mob, or direct democracy. Their leaders were Alexander Hamilton, John Adams, and, later, James Madison, who came over from the popular group. Whereas the Declaration is the main document of the popular group, the Constitution is the enduring work of the strong government group. They were, in their view, giving a practical and necessary answer to the unravelling conditions of the 1780s, when they saw "their" revolution coming unstuck by the purposeful weakness of the barely national government. Both because they saw the weaknesses of government under the Articles and because they were philosophically opposed to much of the anti-institutional democratic spirit of the Enlightenment, they rallied anti-Articles forces in the mid-1780s and caused Congress to summon a convention to meet in Philadelphia in 1787 to revise the Articles. By 1787, there was fairly broad consensus in the new nation that the Articles should be revised in the direction of greater powers for the national (central) government. But—and this is vital—when the delegates to the convention supposed to revise the Articles met in Philadelphia, they decided not to obey their revisionary instructions but to write an entirely new frame of government. Technically this was, as historian Frank Thistlethwaite (1959) writes, a *coup d'état,* or seizure of the state. It was illegal for them to have disobeyed their instructions. They covered this deception by meeting in secret, and the minutes of their deliberations were not made public until many years later.

When one looks at the Constitution, a different style is immediately apparent. Nowhere does one read the "great phrases" of the Enlightenment, speaking of inalienable rights and the pursuit of happiness. Rather, the Constitution is a practical document that sets forth the arrangements for a republican form of government, or, the rule of all by "the best." In a republic, the elite simply do not trust ordinary people to make the right decisions on complex issues of state. While they would never (in theory) use their power for their own good or that of their social class or region, the republican elite would wield power in the name of the people and for their own good. The main emotion out of which the Constitution was written was fear: fear of the one (monarchy) and fear of the many (democracy). For them, life and property could only be secure when it was clear, in John Adams's phrase, that it was "a government of laws, not of men."

Historian Richard Hofstadter (1948) wrote that "the Constitution of the United States was based on the philosophy of Hobbes and the religion of Calvin." This group of founders believed—contrary to the Enlightenment—that the natural state of humans was one of enmity, not amity, toward God and fellow humans. In short, if a government is to write a constitution to govern real people in real circumstances, it ought to take into account the way people really are. And, with a vivid Calvinistic sense of the staying power of evil in human affairs, they drafted a constitution that hoped for the best but erected safeguards against the worst in human potential. The Constitution was an illiberal document in that, through its famous "checks and balances," it saw unordered liberty as a threat to liberty itself.

We conclude this section on the American Revolution by asking if it was an Enlightenment revolution. The answer is yes and no. Yes, it was an Enlightenment revolution in its initial phase, but it reacted against the spirit of the Enlightenment in its ultimate phase. It will never do for us to say "the Founders believed" because there were two sets of founders, arguing different tendencies in political philosophy, issuing in two quite different documents, the Declaration of Independence and the Constitution. In our time, there is an arch-conservative group, the John Birch Society, that has tried to call Americans back to their roots by promoting the following slogan: "This country was founded as a republic, not a democracy; let's keep it that way." In fact, they are wrong: It began as a democracy and became a republic. But I will give them high marks for knowing the difference between the two. Ironically, many "conservatives" in our time, are really democratic liberals in their belief in small government and in people's abilities to make the right decisions if government "got off their necks." We will return to this irony in another connection.

So, will the "real" American Revolution please stand up? In fact, *both* American revolutions live in the memory of Americans, and *both* were "successful" on differing planes of consciousness. On a *rhetorical* level, the American Revolution is a "liberal" revolution, in that people have believed, with Ralph Waldo Emerson, that Americans "fired the shot heard 'round the world," the first shot in the struggle for democratic liberty. On a *real* level, the American Revolution is a "conservative" revolution in that people have believed it preserved liberty under the rule of law. It is the wrong question for us to ask which revolution is "for real" because both exist in the American memory: *Rhetorically* we say America is liberal democracy; *really* we know America is a republic.

The French Case

The French Revolution was, indeed, a revolution. There can be no discussion, as was the case in our consideration of the American Revolution, of whether it qualifies as a revolution. Not only does it qualify, but in a certain way, it is *the* revolution of modern times. Whether or not *the actual* French Revolution of 1789–99 qualifies as an Enlightenment revolution we shall soon take up. But we can say that *the idea* of the French Revolution carries with it a special history (a "myth") that has echoed through subsequent history—from Paris to Moscow to Beijing to Havana to various cities in the developing countries. Moreover, it is not merely a cliché to echo Tocqueville's statement that the revolution was "entirely unforeseen." In truth, it is like one of those great occurrences in human (or our personal) history that could not have been predicted but, when it happened, was seen as an "almost inevitable" occurrence.

France was a nation of paradoxes in the eighteenth century. Unlike Britain, whose Parliament had met regularly for many hundreds of years, the French Estates General had not met since last called in 1614. But, while a nation of little tradition of parliamentary democracy, it was also the nation of the greatest outpouring of the work of the *philosophes*. Despite the work of the *philosophes* and other forces destabilizing the old order, on the surface, the institutions of the old order seemed as secure as ever. Charles Dickens gives the sense of the time, of contradictions surging under an apparently placid and permanent surface, in *Tale of Two Cities:*

It was the best of times, it was the worst of times, it was the age of wisdom, it was the age of foolishness, it was the epoch of belief, it was the epoch of incredulity, it was the season of Light, it was the season of Darkness, it was the spring of hope, it was the winter of despair. . . . There were a king with a large jaw and a queen with a plain face on the throne of England; there were a king with a large jaw and a queen with a plain face on the throne of France. In both countries it was clearer than crystal to the lords of the state, preservers of loaves and fishes, that things in general were settled for ever.

But, alas, things were not "settled for ever." Rather, certain forces would soon unleash themselves that would unsettle things forever.

The French Revolution is another of those major intellectual battlegrounds in modern political thinking. Not only is there greater complexity within the revolution itself—more than in the American Revolution—it is a battleground for the hearts and minds of modern people because interpreters of Marxist persuasion have made it so. They have a great ideological stake in the interpretation of the French Revolution, and that has caused the non-Marxist stake in interpretation to rise correspondingly. So it is difficult to disengage "the myth" of the French Revolution from the revolution itself. For the next few pages we will try to fit the

French Revolution into the concerns of this chapter—the politici-zation of the Enlightenment—and then we can deal with the theme of "myth and reality."

In the American Revolution, we saw two phases in a revolution seen to be successful. In the French Revolution, the story is more complex, with at least four discernable phases in a revolution not necessarily seen to be successful. The first phase of the revolution goes from June 17, 1789, to September 30, 1791. These are key dates because they mark the moving away of the "Third Estate" from the meeting place of the Estates General until the time of the dissolving of the National Constituent Assembly into the Leg-islative Assembly. The goal of the first phase of the revolution was relatively conservative in that what was desired was a constitution-al monarchy on, more or less, the British model in which historic institutions would be maintained but reshaped in the pursuit of moderately "enlightened" goals. In the first phase and in ensuing phases—this is vital—there were "hawks" who wanted to pursue the revolutionary process further and deeper and "doves" who thought the revolution had gone far enough.

The second phase of the revolution—October 1, 1791, to Sep-tember 21, 1792—begins when the revolutionary leadership gives up on the idea of a monarchy of any kind. This year is critical in the life of the new government under the new Legislative Assem-bly because we begin to see real divisions between the factions on how far the revolution ought to proceed. Those wanting radically democratic change sat on the left side of the Assembly and those opposing the radicals sat on the right side (interestingly, we still use the terms *left* and *right* to describe respective positions on change). Since the main texts that this book supplements deal carefully with the facts of the revolution, we need not name all the factions and groups. But, we should mention that the groups on the left in the Assembly ("the Mountain," "the Plain," and "the Girondists") grew impatient with the—in their view—slackening pace of change under the Assembly, and the revolution soon moved to its most explosive, chaotic, and difficult-to-explain phase. In 1789, the Girondists would have been considered radical, but by late 1792 they were conservatives in the renamed frame of government, the National Convention (September 21, 1792, to October 25, 1795). More and more the radicals of "the Mountain" took leadership in respect to two issues: the ever-radicalizing of the revolution itself towards "democracy" and the problem of the militarization of the revolution, because both inside France and from Britain, Holland, Spain, and Prussia there was armed resistance. A revolution begun in a nation-state now became a Europewide revolution, both for those promoting it and for those opposing it. If "Liberty, Equality, and Fraternity" were the goals, why (the radicals asked) should it be limited merely to France, because, as Thomas Jefferson said in the early 1790s, "every lover of liberty has two nations, his own and France."

It is at this third phase of the French Revolution (under the convention, 1792-95) that we should pause for a moment to reflect where we have come and what lies ahead. For our purposes in this book, the third phase of the revolution illustrates both clearly and painfully the most essential political and social question in the modern worldview. The question is this: What is a democracy? Or, how do we know that "the will of the people" is the right thing for society? This is, in short, the problem that Jean-Jacques Rousseau has left us. In the previous chapter we discussed Rousseau and noted his concept of "the general will," which *must* be obeyed. This is *the* essential point of democratic thought, that, as Abraham Lincoln said, "the voice of the people is the voice of God," at least in political terms. In democratic theory, people are free to choose, but they choose well when their choices agree with the general will. Yet, if their free choices reject the wisdom of the general will, the people may have to be forced to be free to do the right thing. In fact, on this view, they will *become* "the people" when, and only when, they agree with the general will. Of course—and here we come to the heart of the matter—this line of thought begs the vital question of who can know the general will, because on such knowledge turns the right and the responsibility of shaping the behavior of the majority in the correct direction. Rousseau, in his time, and all radical democrats since then have no acceptable answer to this dilemma other than to reassert Enlightenment formulas about natural rights and about people's ability to solve their own problems and, in the end, to seize and to use power to force people to be free.

It is this critical time in the French Revolution (1793–94) that later interpreters would cite as the point when the revolution reached its zenith or nadir. In April 1793, the convention established a Committee of Public Safety, which, over the next fourteen months, assumed increasing powers, verging on the dictatorial. The leadership among the radicals at the head of the Committee of Public Safety was Jacques Danton, Lazare Carnot, and, above all, Maximilien Robespierre who, along with Thomas Jefferson in America, was the most famous theorist-turned-revolutionary in the age of the democratic revolutions. In the summer of 1793, the committee's leadership forged a relationship with the semi-organized, radical people of Paris, called the *sansculottes* because of their plain style of dress. This "crowd" is very important in the French Revolution (as the "Sons of Liberty" had been in Boston in 1776), and it showed friend and foe alike what the potential for revolution actually was if hitherto inarticulate people could be politically mobilized. Carnot, with special responsibility for the military side of the revolution, used this mobilization of the lower classes to require military conscription on behalf of the nation and the revolution. This alliance of the lower classes, the idea of the nation, and the revolution caused, more than anything else, the foes of the revolution inside and outside France to organize in reaction.

While a new republic was declared in 1793, its full provisions were not implemented because of the military emergency facing the revolution. Revolutionary tribunals (what today we would call "people's courts") were established in the always more radical Paris but also in the provinces. "Enemies of the revolution" could be denounced, tried, convicted, and executed without much attention to "due process of law." During 1793 and 1794, the policies of the committee came to be known as "the Reign of Terror." The word *terror* is not one given by later historians, but a word used by Robespierre himself to justify "the swift justice" necessary to complete the work of the revolution. But—as all constitutionalists will want to remind Robespierre—when "swift justice" is no longer in judicial but in political hands, who is to determine what is just? In 1794 Robespierre turned on some of his own colleagues in the committee and on the leadership of the *sans-culottes*, and Danton was executed as an "enemy of the revolution."

It is at this vital point that we must pause again and ask—without taking sides in doing so—just what must happen for revolutionaries to think their revolution "complete," or just what must happen for elements in the revolutionary movement to think their revolution has "gone too far"? Obviously this is a highly subjective matter because there is no objective standard with which to judge it. Is it religion? Some revolutionaries remained loyal to the church while wanting to democratize it, as opposed to the radicals' insistence on de-Christianization in 1793, symbolized by the renaming of Notre Dame Cathedral "the Temple of Reason." Even Robespierre thought this was "going too far," and under his leadership a deist civil religion ("the Cult of the Supreme Being") was introduced in 1794. This sort of religion was thought to be revolutionary in previously Catholic France and because it was encouraged, on Enlightenment principles, by revolutionaries. But, in America, a previously Protestant country, such religion appealed to many, if not most, of America's Founders and has not been thought of as revolutionary.

Is *failure* the right word for what happened after the Reign of Terror? Maximilien Robespierre and his followers might have thought so because, in July 1794, Robespierre himself was denounced as an enemy of the revolution and was executed. The "conservative revolutionaries"—the Girondist faction expelled the Convention—returned with considerable organizational skill to redirect the course of the revolution. In October 1795, the convention was replaced by the Directory, and until its demise in 1799, it swung the pendulum back toward the right. Terror, in its former guise, ended, although many of the radical and *sans-culottes* leadership in Paris were victims of a counterterror. The leadership in the Directory was drawn from relatively moderate elements in the middle classes. Had there been no external wars to fight, these middle-class leaders might have brought the revolution to rest in something like the manner of the American Revolution. But with France's military situation ever worsening, the Directory turned more and more to the military for leadership. With the rise of Napoleon Bonaparte, the Directory conflated into the Consulate (1799) and the Empire (1804), with Napoleon crowning himself "Emperor of the French." In a long and bitter series of battles, the British, Prussian, and Austrian forces were to defeat the French, the culminating defeat coming at Waterloo in 1815.

Was the French Revolution a "failure"? On the terms argued in this chapter, i.e., was it an Enlightenment revolution, the answer must await that to another question: Which French Revolution do you mean? If "the real" French Revolution be the principles of 1789 (constitutional monarchy) or the principles of 1792 (democratic republic), then the French Revolution is a moderate success because it did sweep away forever the apparatus of the *ancièn regime*. But, if "the real" French Revolution be the principles of 1793 (direct democracy, secular society) then it must be regarded as a failure because it did not bring into being the new person in a new society envisioned by the *philosophes*. On these questions the debate between non-Marxists and Marxists continues to turn.

Evaluating the Age of Revolutions

In conclusion, we can say a few things about the age of democratic revolutions. The "old order" of hereditary monarchy, fixed social classes, and a privileged position for any Christian church (whether Protestant or Catholic) was challenged and, in the end, supplanted. The new arrangements varied from nation to nation, but it was certain that the old order was lost indeed. For many of the *philosophes* that is victory enough. But, for the more radical *philosophes,* the mere dismantling of the old order without replacing it with a liberal democracy represented an opportunity lost and one to be sought again at the first possible moment. So, in the end, the debate about the American and French revolutions is not only about the past but also about the future. The stories of "the rising" of the American and French peoples in the late eighteenth century continues to provide powerful "myths," both to citizens of the two nations and to other later peoples who, believing themselves oppressed, look to Philadelphia and Boston (1776) and, pre-eminently, Paris (1789) for inspiration.

Finally, what can be concluded on Christian terms? We must be careful of labels here, remembering that terms like *conservative* and *liberal* cloud more than illumine the argument. It would seem consistent with what we have formerly said (i.e., that Christianity and the Enlightenment are theoretically incompatible) to suggest that Christians might observe the failure of Enlightenment principles and say, "That's to be expected." But, the question remains: Does that conclusion lead Christians to observe that the American and French revolutions were, in the end, middle-class revolutions and say, "That's good"? In short, it is all very well to oppose the radical version of the Enlightenment on Christian grounds (as the Dutch politician Abraham Kuyper later said, to be "antirevolutionary"). But Christians must beware lest they back into a baptizing of middle-class republics. And this is all the more clouded by the fact that, in the United States, many Christian citizens confound the clarity of the matter by the very baptizing of a middle-class republic in terms *not* of the *reality* of a republic but the *rhetoric* of the democratic Enlightenment. They call conservative the radical ideals of the eighteenth century—individualism, opposition to central government, and the belief that a good society offers scope for an ever-expanding liberty. The Constitution of the

United States was, after all, "based on the philosophy of Hobbes and the religion of Calvin." It presupposes a view of society that is historically illiberal, and, one supposes, Christians today who still believe in that way must be corporatist and elitist, not individualist and democratic. For Christians to confuse the *rhetoric* for the *reality* of the age of revolutions and to champion the liberal values of the Enlightenment while calling them conservative values consistent with a Christian worldview is to join Lord Cornwallis's band in playing the song "The World Turned Upside Down."

Chapter 12

THE INDUSTRIAL REVOLUTION:
THE TEST OF A WORLDVIEW

The industrial revolution is one of those great sets of massive social changes that causes us to value the term *revolution*. And, like Tocqueville on the French Revolution, we might say that its results were vast and not entirely foreseen. For us to "cover" the industrial revolution is more of an undertaking than this book is prepared to do. For our purposes, the industrial revolution is a very interesting test of the working out of the Enlightenment worldview because we will look into the large-scale changes we associate with the industrial revolution and then inquire into whether or not it did, or did not, do what its champions said it did. In short, we shift from the sociopolitical analysis of the last chapter to socioeconomic in this one, but the questions are similar: What were the main engines of change in the modern West, and did life actually get better, as the Enlightenment promised it would? These are important questions, both because the answers to them are vital for our understanding of contemporary society and because, in discussing the main responses to industrialization, we discover the origins of modern politics. In this chapter we must first gauge the phenomenal changes that began in, say, 1740 and then observe that Enlightenment ideas like liberty and equality are definitely linked to explanations for industrial society, known alternatively, as capitalist and socialist.

The Scope of Economic Modernization

We value the term *industrial revolution* because it suggests the massiveness of the economic changes. In fact, however, we now think that the industrial revolution was a phase in a much larger and longer process, now increasingly called by scholars *economic modernization*. In developing the theory and practice of economic modernization we shall rely on the work of Robert L. Heilbroner (1980) and Walt W. Rostow (1960). The roots of the great economic changes we will discuss lay deep in Western history and go back to the dissolving of the Middle Ages into the new spirit of the Renaissance, discussed several chapters ago. In sum, in seeing the transition from traditional society to modern society we see a process of development from agricultural to industrial, from simple to complex, from rural to urban, from local to national and international.

A traditional economic society is well described by Rostow. It is a society "whose structure is developed within limited production functions, based on pre-Newtonian science and technology, and on pre-Newtonian attitudes toward the physical world. Newton is here used as a symbol for that watershed in history when men came widely to believe that the external world was subject to a few knowable laws and was systematically capable of 'productive manipulation.'" Now in our time the word *manipulation* carries a negative, especially psychologically negative, connotation. But it need not necessarily be so. To the people who acted on Enlightenment principles, manipulation merely meant that, through the knowledge of natural laws, humans could begin to shape their own world rather than be shaped by it. In fact, it is in the emergence out of traditionality that we see the full convergence of economic, political, religious, and social ideas and forces. Therefore, it comes as no surprise to us that Britain was the first nation to modernize economically because, among other reasons, it was the first nation to stabilize its national political structures in terms of humanist-Protestant-Enlightenment ideas. In order for a society to modernize economically there are, apparently, preconditions that demand a prior acceptance of an entire worldview that, not only is the world orderly and predictable, but that it is knowable and controllable by humans.

While the details may vary from nation to nation, all modernizing societies must go through something of the same process, or, in the terms of this book, of religious disenchantment, of humanist assertiveness, and of a growing confidence in human manipulative potential in respect to a "rational" world. Robert Heilbroner rightly stresses that what is necessary for economic modernization is the concurrent emergence of several structural and attitudinal changes. If we look into these necessary changes we can see why Britain modernized first and why its experience of the industrialization process offers a typology for the modernization of other nations.

Britain was the first nation not only to create substantial wealth but to distribute it fairly broadly. It was not only the nobility that possessed wealth in the eighteenth century; there was substantial wealth possessed by an upper-middle group of people that, functionally, can be called a *commercial bourgeoisie*. Although on a smaller scale by later standards, Britain was first among nations in creating a consumer market, and, as a result, the rising demand for consumer goods and services fueled the fires of the search for new techniques to supply those goods and services. Moreover, Britain was the first to break the power of local nobility and to create a national citizenship in a single, unified state. This, again, is a strong incentive to a rising mercantile class. With the expelling of small farmers and agricultural laborers from the countryside through the "enclosure" movement, in which land lords "enclosed" small holdings into larger estates—from small-scale subsistence agriculture to large-scale capitalist agriculture—a large number of laborers were available to become the work force for a nascent industry.

What is particularly interesting, for the purposes of this book, is that Britain was a center of interest in technology, i.e., science and engineering. The Royal Academy, founded in 1660, with Isaac Newton himself as one of the first presidents, helped to create a broad interest in machines and especially in new inventions that would "improve" something. Whether the improvement was in agriculture, mining, transportation, or manufacturing, British leaders were interested in them in a way that distinguished them from, say, their French counterparts. For reasons not altogether provable, British landowners, for example, were interested in schemes of "scientific" agriculture, even down to fertilizers, that French landowners would have thought beneath their dignity.

More than anything else, however—and this is vital—the British developed first what Heilbroner calls "new men." These people of the expanding commercial bourgeoisie were both self-consciously innovators, in that they eschewed the old and sought the new, and self-consciously entrepreneurs, in that they were always on the lookout for an opportunity to invest, expand, and develop their economic interests. As Heilbroner rightly points out, the creation of wealth for the "new men" was incidental to their real aim, the creation of power: "In an economic, if not a political, sense, they deserve the epithet 'revolutionaries,' for the change they ushered in was nothing short of total, sweeping and irreversible."

Religion and Capitalism

In our attempt to understand this sweeping economic change to a market economy, we must pause for a moment to consider the role of religion in the rise of market capitalism. The twentieth-century historical sociologist Max Weber gave the classic formulation of religion's role in his influential book *The Protestant Ethic and the Spirit of Capitalism*. While many of Weber's views have been debated and challenged, we can, nevertheless, see his point: that Protestant nations forged ahead in the economic race, and, insofar as Catholic nations also developed, they did so in a "Protestant" way. What is the Protestant way of business? In contrast to Catholic socioeconomic thinking, which tended to stress an "other worldly" approach to the mundane aspects of life, Protestant thinking was "world affirming." From two major strains of Protestant theology—the Lutheran and the Calvinist—came the idea of "calling," in which simple distinctions between secular and sacred were abolished and in which all occupations and endeavors were no more or less "kingdom" service. Above all, the Calvinist worldview was one that urged a life of rectitude and diligence. This provided a powerful stimulus to the economic lives of the so-called new men. While no ideologist can be cited as having said that a large bank account (or other measure of economic success) was an index of God's blessing on a particular individual, there was an inferential implication that that was so. In fact, if a person had some capital, and if he did not think mundane work to be beneath him—to the contrary, if he thought himself "called" by God to it—that person *is* likely to be successful. Not only does the so-called Protestant work ethic have impact on the creation of wealth, it says a great deal about the use of wealth. The work ethic encourages people to work, but they were not to indulge themselves with the profits but, rather, to use the profits (as "stewards") to create more wealth and, incidentally, livelihoods for others. This was done through an idea—commonplace to us, but new in the eighteenth century—of *thrift*. Thus, investment (the use of savings for a productive purpose) became an index both of a person's economic good sense and of his religious worth. As Heilbroner concludes, "Calvinism fostered a new conception of economic life. In place of the old ideal of social and economic stability, of knowing and keeping one's 'place,' it brought responsibility to an ideal of struggle, of material improvement, of economic growth." While this view of life is not the sole preserve of Protestants—some recent studies of Catholic businesspeople in Amsterdam show that they were more "Calvinist" than the Calvinists in this respect—the historic role of the so-called Protestant work ethic was that of a legitimating ideology, in which the new men of the new order could find a religious framework for their secular success.

There *was* enormous success for those nations that industrialized and modernized. Over the long run and in macroeconomic perspective industrial capitalist nations provided goods and services to their people that were, in the early eighteenth century, unimaginable except to the very wealthy. There is no doubt that, in the end, everyone in an industrializing nation benefitted to some degree. Life grew quantitatively better, but whether or not qualitatively better is a matter we shall return to later in this chapter. But to say that everyone benefitted *in the end* is not to say enough, because the end was a long time coming for many people, and—importantly—not within the range of their life spans. And to say that macroeconomically there was measurable improvement is, again, not to say enough because, within any economic system, the distribution of goods and services was markedly unequal. The fact that the "success" of industrial capitalism was—and is—accompanied by an increasing inequality is enough for the moral debate of capitalism to have begun in the initial phase of industrialization and to continue to our own time. While there was a world of pleasure for some, there was a world of pain for many as the process of the industrial revolution went ahead. We will return to Karl Marx later in this chapter, but for now one matter associated with his name needs to be cleared away. He asserted that, in the industrializing process, "the rich got richer and the poor got poorer." In this, Marx is half wrong and half right: He is wrong in that the poor did *not* get poorer but got a bit better off while the rich got very much richer; he is right in that the gap between the rich and the poor grew greater. Even in the United States, where industrial success was greater than in any other nation and where distribution has been relatively more equal, nevertheless the wealth of the top 10 percent continually has increased considerably faster than that of the bottom 50 percent. So throughout American history, while most people did better much of the time, the maldistribution of wealth grew greater all the time.

The "Worlds of Pain"

Never mind the long run, in the short run—in the initial stages of industrialization—the majority of laborers were not advancing at all but sinking into degraded and wretched conditions. As John Masefield (1878–1967), the late poet-laureate of Britain, wrote in "The Everlasting Mercy,"

> To get the whole world out of bed
> And washed, and dressed, and warmed, and fed,
> To work, and back to bed again,
> Believe me, Saul, costs worlds of pain.

We have some clear picture of what the "worlds of pain" were like because of the activities of some committed Christians in Britain. During the first phases of the industrializing process—when working conditions for men, women, and children were most brutal—many leading Christians reacted in horror to the appalling conditions of life and labor, and they demanded political action on behalf of the poor. Most notable among them is Anthony Ashley Cooper (1802–85), better known as the Earl of Shaftesbury. Despite his privileged position in the nobility, Shaftesbury's life and work is a clear testimony that Christian commitments override class positions and that the Christian gospel's preference for the poor is not mere metaphor. Along with his colleague in Parliament Michael Sadler (1780–1835), he pressed for, and got, a commitment from Parliament to investigate the conditions of work in the newly developing industries. The report of the parliamentary commission, called the *Report from the Committee on the Bill to Regulate the Labour of Children* (1832), is commonly known as the Sadler Report. The report is not an analysis of working conditions but the *verbatim* testimonies of many persons at work in factories and mines. The life of all workers was appalling, and all the more so for children. An excerpt from the Sadler Report is no less shocking now than it was then.

THOMAS BENNETT, called in; and Examined

Where do you reside?———*At Dewsbury.*

What is your business?———*A slubber.*

What age are you?———*About 48.*

Have you had much experience regarding the working of children in factories?———*Yes, about twenty-seven years.*

Have you a family?———*Yes, eight children.*

Have any of them gone to factories?———*All.*

At what age?———*The first went at six years of age.*

To whose mill?———*To Mr. Halliley's, to piece for myself.*

What hours did you work at that mill?———*We have wrought from 4 to 9, from 4 to 10, and from 5 to 9, and from 5 to 10.*

What sort of a mill was it?———*It was a blanket-mill; we sometimes altered the time, according as the days increased and decreased.*

What were your regular hours?———*Our regular hours when we were not so throng, was from 6 to 7.*

And when you were the throngest, what were your hours then?———*From 5 to 9, and from 5 to 10, and from 4 to 9.*

Seventeen hours?———*Yes.*

What intervals for meals had the children at that period?———*Two hours; an hour for breakfast, and an hour for dinner.*

Did they always allow two hours for meals at Mr. Halliley's?———*Yes, it was allowed, but the children did not get it, for they had business to do at that time, such as fettling and cleaning the machinery.*

But they did not stop in at that time, did they?———*They all had their share of the cleaning and other work to do.*

That is, they were cleaning the machinery?——*Cleaning the machinery at the time of dinner.*

How long a time together have you known those excessive hours to continue?——*I have wrought so myself very nearly two years together.*

Were your children working under you then?——*Yes, two of them.*

State the effect upon your children.——*Of a morning when they have been so fast asleep that I have had to go up stairs and lift them out of bed, and have heard them crying with the feelings of a parent; I have been much affected by it.*

Were not they much fatigued at the termination of such a day's labour as that?——*Yes; many a time I have seen their hands moving while they have been nodding, almost asleep; they have been doing their business almost mechanically.*

While they have been almost asleep, they have attempted to work?——*Yes; and they have missed the carding and spoiled the thread, when we have had to beat them for it.*

Could they have done their work towards the termination of such a long day's labour, if they had not been chastised to it?——*No.*

You do not think that they could have kept awake or up to their work till the seventeenth hour, without being chastised?——*No.*

Will you state what effect it had upon your children at the end of their day's work?——*At the end of their day's work, when they have come home, instead of taking their victuals, they have dropped asleep with the victuals in their hands; and sometimes when we have sent them to bed with a little bread or something to eat in their hand, I have found it in their bed the next morning.*

Had it affected their health?——*I cannot say much of that; they were very hearty children.*

Do you live at a distance from the mill?——*Half a mile.*

Did your children feel a difficulty in getting home?——*Yes, I have had to carry the lesser child on my back, and it has been asleep when I got home.*

Did these hours of labour fatigue you?——*Yes, they fatigued me to that excess, that in divine worship I have not been able to stand according to order; I have sat to worship.*

So that even during the Sunday you have felt fatigue from your labour in the week?——*Yes, we felt it, and always took as much rest as we could.*

Were you compelled to beat your own children, in order to make them keep up with the machine?——*Yes, that was forced upon us, or we could not have done the work; I have struck them often, though I felt as a parent.*

If the children had not been your own, you would have chastised them still more severely?——*Yes.*

What did you beat them with?——*A strap sometimes, and when I have seen my work spoiled, with the roller.*

Was the work always worse done at the end of the day?——*That was the greatest danger.*

Do you conceive it possible that the children could do their work well at the end of such a day's labour as that?——*No.*

Did not this beating go on principally at the latter part of the day?——*Yes.*

Was it not also dangerous for the children to move about those mills when they became so drowsy and fatigued?——*Yes, especially by lamp-light.*

Do the accidents principally occur at the later end of those long days of labour?——*Yes, I believe mostly so.*

Do you know of any that have happened?——*I know of one; it was at Mr. Wood's mill; part of the machinery caught a lass who had been drowsy and asleep, and the strap which ran close by her catched her at about her middle, and bore her to the ceiling, and down she came, and her neck appeared to be broken, and the slubber ran up to her and pulled her neck, and I carried her to the doctor myself.*

Did she get well?—— *Yes, she came about again.*

What time was that?—— *In the evening.*

You say that you have eight children who have gone to the factories?——*Yes.*

There has been no opportunity for you to send them to a day school?——*No; one boy had about twelve months' schooling.*

Have they gone to Sunday-schools?——*Yes.*

Can any of them write?——*Not one.*

They do not teach writing at Sunday-schools?——*No; it is objected to, I believe.*

So that none of your children can write?——*No.*

What would be the effect of a proper limitation of the hours of labour upon the conduct of the rising generation?——*I believe it would have a very happy effect in regard to correcting their morals; for I believe there is a deal of evil that takes place in one or other in consequence of those long hours.*

Is it your opinion that they would then have an opportunity of attending night-schools?——*Yes, I have often regretted, while working those long hours, that I could not get my children there.*

Is it your belief that if they were better instructed, they would be happier and better members of society?——*Yes, I believe so.*

The Origins of Modern Politics

The response to, and interpretation of, economic modernization proceeds along two lines of thought, historically known as liberal and egalitarian. Even though liberals would later, and only in the United States, call themselves conservatives, and even though egalitarians would later, and only in the United States, call themselves liberals, it is best for us to use the historical names for purposes of clarity. Moreover, it is important to note that there is no longer a historic conservative position. Historical conservatives are those who sought to preserve the *ancièn régime,* i.e., an "old order" of monarchy, fixed social classes, and established religion. But, in the age of the democratic revolutions, the structures of the old order were replaced, as we have already seen, with the modern administrative state. In the nineteenth and twentieth centuries *all* systematic thinking on socioeconomic-political matters has gone on within an Enlightenment framework.

<u>EQUALITY</u>   <u>LIBERTY</u>
<u>ENLIGHTENMENT</u>

As this simple diagram shows, *all* modern thinking stands on an Enlightenment base. Whether liberal or egalitarian, all the main writers share the Enlightenment penchant for seeing things scientifically, for expecting progress, and for believing that humankind can alter its own future. The liberals and the egalitarians alike believe in the improvement of society—even the "perfection" of society—and what separates them is not their overriding worldview concerns but the methods by which the expected and hoped for progress of humankind will proceed. Let us take, in turn, the main writers who spoke for a liberal view of society (especially Adam Smith) and then the main spokesman for egalitarianism, Karl Marx.

Adam Smith is the main spokesman for liberal economic theory, though we will want to make mention of others, such as Thomas Malthus and David Ricardo. Taken together, their view of economic life goes under the heading of classical economics. Readers of this book will, by now, know what is "classic" about classical thinking, i.e., the application of reason. So, we see classical economics as firmly rooted in the rationalism of the Enlightenment. Smith's magisterial work, for which he is much remembered and frequently quoted, is *An Inquiry Into the Nature and Cause of the Wealth of Nations* (1776), though it is more popularly known by the shorter title *The Wealth of Nations.* We cannot overemphasize the importance of Smith's work because it is the classic statement of liberal economics and the basic primer for those who, two hundred years later, call themselves conservatives.

Smith is the classic liberal. He believes in freedom. At the time of his writing, in Britain the economy was regulated by a set of arrangements known as mercantilism. He suggested that these various laws and regulations be abolished because they were "unnatural" (a key Enlightenment word), and he advocated a "natural" system of economic liberty. He believed that economic growth, and therefore, the wealth of any nation, would only happen if individuals were loosened from the unnatural bounds (or, breaking "the chains" of Rousseau). People could pursue their individual self-interests both as producers, by meeting the demands of a market economy, and as consumers, by freely choosing the ways their wants might be met. This pattern of economic thinking is often called *laissez-faire* (literally, "to be allowed to make") or *laissez-aller* ("let it go") because the best society, it is said, would emerge if individuals were allowed to make and to go where their own interests dictated in the working of a free-market economy. Since, in this liberal view of society, the individual is the focal point, government should be minimally interventionist and the individual should be maximally free to choose. Moreover, the boundlessness of human potential is analogous to the boundlessness of nature ("Nature") itself. In Smith, the natural resources of the world are placed there by "Nature's god," and people ought to use them for the betterment of humankind.

The seeming innocent and naive qualities of Smith's hope for the future can perhaps be explained by his time of writing, in 1776. The worst facts of factory and mine work had not yet become common knowledge. Yet as laissez-faire liberals in the tradition of Smith wrote in the later context of industrialization, they did not change their attitudes on state intervention, and their writings show a harsh quality absent from the more genteel Smith. Thomas Malthus (1766–1834), in his famous *Essay of Population* (1798), argued that the "natural law" of population was that people multiply while food sources add (or, as he put it in the mathematical language of the Enlightenment, populations grow geometrically while food expands arithmetically). Moreover, there was no way to avoid the disaster of famine, disease, and poverty because, he said, if the state intervenes to help the poor, they will only have more children, thus causing an even greater gap between population and food. David Ricardo (1772–1823) went even further than Malthus. In his *Principles of Political Economy* (1817) he agreed with Malthus but raised the argument to the level of natural law, what he called the "Iron Law of Wages," which relates wages to "surplus" population: In short, if wages are kept low, fewer children will be produced to put undue demands on food.

The main liberals—Smith, Malthus, and Ricardo—gave a picture of economic life under capitalism that has informed political and economic thinking up to our own time. There are, they said, certain unchangeable laws of nature that function as implacably in the economic world as gravity does in the natural world. If this meant a period of hardship for some, that hardship would only be for the interim. In the end, everyone would be better off if we let the natural laws do their inexorable work and if we allowed those capable of creating wealth to do their entrepreneurial work. In short, if government would stop impeding progress by restricting and regulating individual action, the best for all would ensue because even the poorest element in the society would be taken up by the general rise in "the wealth of nations."

Workers in the new factories of the cities and in the mines hotly disputed this scheme of things. Their representatives typically did not arise from the distressed classes but were middle-class people who had insight into, and empathy for, (what Marx's friend Friedrich Engels called) "the condition of the working class in England." That condition, as we have already seen in the Sadler Report, was ghastly for many, even desperate for some. In general, critics of existing social arrangements wrote on what they called "the social question," and for them the alternative to "individualism" was "socialism." It is important for readers from North America, where there has not been a viable socialist movement, to realize that in Europe socialism does not carry with it the negative connotation it does here. Critics who began focusing on "the social question" about 150 years ago had no political base. But, in the intervening period, socialist parties have become an essential part of the political fabric in Europe. Let us look at a few important socialist writers before giving an extensive analysis of socialism's main spokesperson.

Egalitarians—like their liberal counterparts—took for granted the normative correctness of the Enlightenment. An early writer, Claude Henri (1760–1825), better known by his noble title, the Count of Saint-Simon, had been a supporter of both the American and French revolutions. He believed in an organic society, i.e., one in which there was a fundamental connectedness between things. In his view, the organic unity present in the Middle Ages had disintegrated during the Renaissance-Reformation period and had further fragmented during the age of democratic revolutions and the early stages of industrialization. The reintegration of industrial society was possible, he thought, if Enlightenment principles would be thoroughly applied. For him, it was not so much a matter of private ownership versus public ownership but of seeing society as it is—a whole—and of managing society's natural resources and economic structures for the good of all.

Another important egalitarian thinker-activist was Robert Owen (1771–1858), who clearly reflects Enlightenment values in his many writings, most notably *The Crisis: Or, The Change From Error and Misery to Truth and Happiness* (1832). Like many Enlightenment thinkers, of both left and right, Owen was a convinced environmental determinist (i.e., he believed that if the conditions of life and labor improved, so would people's character). This was not a mere visionary scheme for Owen, who built industrial communities in both Scotland and in Indiana, the latter significantly named "New Harmony." These experimental communities flourished for a time, but they did not live much longer than the period of the original enthusiasm of their founder. Nevertheless, Owen's contribution to egalitarian thought was that cooperative behavior could also be practical and profitable in an industrial setting.

Karl Marx is the most important thinker in the egalitarian tradition of Enlightenment thought. His continuing importance is founded on the reality that about one-third of humanity in our time live in nations that claim some direct lineage from Marx's inspiration. For our purposes the importance of Marx is not what has been made of him in the twentieth century but what he wrote about "the social question." Yet, we should admit the difficulty in separating the historical Marx from the mythic Marx of the Marxist movement. V. I. Lenin—the Soviet Union's founding father—wrote that there were three influences upon Marx's work: the philosophy of Hegel, from whom he borrowed the idea of the dialectic, though changing it from the realm of ideas to that of material reality; the classical economic thinking of Adam Smith, from whom he learned the labor theory of value, though changing it from a defense of capitalism to an indictment of it; the socialist writings of Saint-Simon and others, from whom he learned about organic society, though transforming what he saw as "utopian" socialism in their writings to the "scientific" socialism of his own. While Lenin is right about these influences on Marx, there is another influence that neither Lenin nor the ideologists of the Marxist movement would either recognize or affirm: the tradition of moral protest stemming from the Judeo-Christian character of Western thinking. Whether or not one approves of or uses Marx, it is indisputable that his initial—and lasting—impact was due, in large part, to the moral outrage in his writings about the degradation of humans in the industrializing process.

Of all Marx's important points—and the texts that this book supplements discuss them—the most important for our purposes is his idea of *alienation*. This is vital both because in it we see Marx most clearly as an Enlightenment man and in it we see the reason for Marx's continued attractiveness. The essence of human nature for Marx is not that humankind was created but that humans create. The essential difference between animals and humans was, as Marx wrote in *The German Ideology*, that humans "begin to produce their means of subsistence, a step which is conditioned by their physical organization. By producing their means of subsistence men are actually producing their actual material life." In Marx's thought, then, we have a radically secular world in which humankind is answerable only to itself. Humans *(homo faber)* make their own world, and, if the structures of that making ("the modes of production") could be reformed in terms of equality, then humankind could be saved. Here we see Marx, although unambiguously a secular man of the Enlightenment, at his most insightful and morally attractive presentation.

A person's essence is lost, Marx insists, when he or she loses control of personal destiny. If the modes of production in capitalism justify—as they did—the treatment of persons as commodities, the person's humanity is lost when work is taken apart from the being of a whole person. The product of work, which ought to be the expression of human worth, is in capitalism treated as an object outside the worker. Thus, both product and worker became commodities and people—at their most essential— are *alienated* from the things they create and from themselves. Further, the work that most people do is forced on them and done for another person, thus robbing them of the spontaneous enjoyment of work, the thing that makes them distinctively human. And, worse still, a person cannot produce in a social way (what Marx called "for species-beings"), which results in the ultimate alienation, of person from person. Again Marx: "In general, the statement that man is alienated from his species-being means that one man is alienated from another as each of them is alienated from the human essence."

Marx's alternative in his envisioned communistic society was that work could be both enjoyable and social and thus protect people's essential humanity. Whether or not Marx's alternative is feasible, and there is a continuing debate on that point, his great contribution in describing industrial capitalism is in pointing out the alienative effects of work in the kinds of situations we noted above in the Sadler Report. Beyond the normative qualities, or lack thereof, of Marx's analysis, many people in industrializing societies in the century and a half since his writing believe that he describes the reality of their experiences. This accounts for the continuing appeal of Marxism for people who themselves cannot understand the intricacies of, say, dialectical materialism, but who, when told about alienation, affirm that in Marx they have found someone who understands them and their situation.

As Christians try to analyze Marx, the two best books to help them through the ideological divisiveness of writings about Marx and Marxism are the objective and fair work of David McLellan (1974) and the insightful analysis from a Christian perspective by David Lyon (1979). As Lyon writes, while there may be important instances in which Christians can learn from Marx, that learning must not forget the very different starting points of Christian and Marxian analyses:

The Christian has to admit that worker/product, worker/self and worker/fellow alienations do occur. He cannot deny that at times Marx came embarrassingly close to a biblical understanding of cooperative, satisfying and unalienated work. But the Christian can never concede Marx's devastating disclaimer—his rejection of the Creator. For the Christian conception of the person, as Marx well knew, begins with God, not people.

We have seen in this chapter that the origins of modern politics lie in the responses to the industrial revolution. Historic liberals (roughly, today's conservatives) believe in liberty for individuals and believe that when individuals are freed from unnatural restraints, and live by natural laws, the best and most successful society emerges. Historic egalitarians (roughly, today's liberals) see society in terms of the relationships of persons and believe that when persons are harmoniously related, as nature intended, the best and most successful society emerges.

What is common to both liberal-capitalist and egalitarian-socialist thinking is a prior acceptance of the Enlightenment worldview, with its stress on science, rationality, natural law, and—above all—progress. For Christians, this line of thought is not nearly good enough. The Enlightenment, as we have observed, provides the basis for modern, secular thinking and behaving. It is a "secular religion" that, in substantial senses, intends to supplant the dominant religion in the West, Christianity. Its cunning attractiveness, as we observed in chapter 10, is that the Enlightenment secular religion is a kind of Christian heresy in which historic Christianity's expectation for a perfect life in heaven is secularized and promised on earth. The "heretical" qualities of the Enlightenment are also seen in its perversion of the Christian notion of providence into the certainties of natural law.

It is not a desire to escape realities that causes us to say with conviction that *no* economic and social system based on the Enlightenment is acceptable to Christians. Nor is it some cute, academic plea for moderation between capitalism and socialism to say that *neither* system—as we have known them—is acceptable to Christians. Both are materialist systems and are, therefore, *inherently* unacceptable.

It is sometimes said, especially in North America, that Christians must put away pious theories and live in the "real" world. But, what passes for the "real" world must be called by its actual name—a fallen world. The way that things have worked out in so-called reality is not the way the Lord of Lords intended them to work out. As Christians in the last years of the twentieth century look to the future of their economic systems, they will do well to realize that all systems of political economy—capitalist, socialist, or "mixed"—are founded on, and justified by, an overriding Enlightenment ideology in which the Christian can find no legitimate place. If Christians were ever to bring a Christian system into being, i.e., one that guarantees community while respecting individuals, that would surely be a remarkable industrial revolution.

Chapter 13

THE IDEA OF PROGRESS: MODERNITY TRIUMPHANT

Historian Peter Gay (1966) sums up Enlightenment attitudes by quoting the German poet, Johann Wolfgang von Goethe (1749-1832), who said in 1792, "Here and today begins a new age in the history of the world. Some day you will be able to say—I was present at its birth." As we have seen, many people were impatient to get on with life in the new age of possibilities, and few writers expressed that sentiment as well as English poet William Wordsworth (1790-1850), "Bliss was it in that dawn to be alive, but to be young was very heaven." I suppose we need not take Wordsworth's invocation of heaven too literally, but it does point to a direction of thought common in the nineteenth century. The future was to be welcomed eagerly and, if not exactly heavenly, it was to be a marked improvement over the past. In short, progress was thought to be inevitable: It *would* happen—of that there was no doubt—as Nature worked its inexorable work, and it would result in the continual increase of the happiness of humankind.

Faith in progress is one of the hallmarks of modern thought, and especially in the nineteenth century it was held not only by intellectuals, but, as we shall see, ordinary people as well believed in it. It is this faith in progress, however, that marks us off from the nineteenth century. Even though the nineteenth century is still quite recent, and even though some people in North America still cling to Enlightenment propositions, most of us sense—almost more than we know—that the idea of progress is more rhetoric than reality except in simplistically material terms. For us, the experience of several world wars, the attempted genocide of European Jewry, and the development of "doomsday" weapons has called the notion of continued human progress into serious question. As historian and economist Robert Heilbroner (1961) has suggested, once in Western thinking the future had a history; the past and the future could, and would, be linked through the triumph of progress. Few serious observers in our time would echo the belief of the German mathematician and philosopher of the eighteenth century Gottfried Leibniz that "this is the best of all possible worlds because all grows better and better every day." If the world is getting better every day it escapes our notice as we endure political and economic instability, terror, and famine.

The question we will pursue in this chapter is this: How could people in the nineteenth century have really believed in progress? Put another way, since our experience does not validate continued human betterment, how can we understand our immediate forebears in the nineteenth century for whom a belief in progress was an essential element of their worldview? In order to understand them—and ourselves in the process—we will have to engage in a considerable leap in historical imagination to see life as they saw it and experienced it. Hope for the future was, after all, one of the hallmarks of modern (Enlightenment) thinking, and, even if we have to give it up in our time, it is important to realize the men and women in the nineteenth century were not fools to believe in human progress. So, let us look at nineteenth century ideas under three headings: peace, technology and science, and America.

The Nineteenth Century Was a Century of World Peace

One of the greatest difficulties for Christians in surveying the history of Western civilization is the realization that despite Christianity's formative influence, the history of the West has been one of nearly continuous warfare. Indeed, in many instances Christianity was at the center of warfare, and, in any case, it has apparently done little to restrain war. The advent of Protestantism in the sixteenth century did not change things in this respect and even contributed to the wars through the seventeenth century. Protestantism often provided a legitimating ideology for the wars of nation-states in later centuries (for example, "Manifest Destiny" in nineteenth century America).

The truly remarkable thing about the nineteenth century was the relative absence of war. The key word, of course, is *relative* because there were some wars. But, relative to what had gone before and what came after, the nineteenth century was one of peace. Between 1815, the end of the wars related to the French Revolution, and 1914, when "the guns of August" began the terror of the twentieth century, there was a period of relative calm. There was no general or, as we would say, "world" war between 1815 and 1914. To be sure, British, French, and Russian forces fought in the Crimea, and the French and Germans fought in the Rhineland. But these wars were short-lived and did not bring in a large number of other nations. Then, there was the American Civil War, which, though a considerable matter for Americans, did not involve other nations. Also, there were continual wars against Indians and the invasion of Mexico in 1846. So, while these rather restricted wars did happen in the nineteenth century, the generalization still holds, that the century between 1815 and 1914 was one in which warfare did not occupy center stage and did not dominate the lives of people in Western society.

The removal of the scourge of war from the consciousness of Western people provides an important clue as we witness the extraordinary amount of energy expended in productive tasks. When a nation is at peace, it often does not dwell on the absence of war, but it gets on with other tasks. Perhaps an analogy will help. When we have a toothache, it seems we can think of little else but our discomfort, and all other activities are abandoned or reduced until the offending tooth is fixed. But once the tooth is restored to health, we rarely think about it. We get on with our lives, and, moreover, we use our teeth for their intended purpose. (Sitting in the pizza parlor with our friends, we do not stop and say, "Gee, I don't have a toothache tonight!" No, we eat the pizza and enjoy the company of friends.)

It seems as though there is a finite amount of energy that any society, or person, has. If there is a constant, unproductive drain on energy, it limits the amount that can be applied to productive purposes. Back to our toothache analogy for a moment. If the ache is caused by an abscess, then the tooth is diseased, and the body can become generally diseased if the abscess breaks open. So, it is not merely a psychological problem to have a toothache; it can be a physical one. War is like a disease, and its infection spreads throughout a society as it becomes more general.

Those of us who have experienced war directly, or who have been a noncombatant adult during wartime, know this to be true. The present writer came to adulthood during the longest American war—Vietnam. My generation of students did the same sorts of things that all North American young people did and continue to do: We made plans for our future, went to university or college, associated with friends, watched baseball or hockey games, borrowed our parents' car, and, generally, got on with life. But, with military conscription ("the draft") a fact of our lives, we had to plan, even during peacetime, to give two or more years to national service. As American commitment in Vietnam deepened during the 1960s, those of us of the ages appropriate for service found ourselves with the curious and recurring interference in our lives of the reality of having to go to war. For us, the war was not a subject to be discussed but a present reality, and few evenings together with friends ended without some evocation of the fright we felt. Further, as the various ones of us were drafted, or volunteered (in exchange for the promise of special training), we kept in touch with each other's families to find out how the friends in the war zone were doing. When the news came of battles in certain parts of Vietnam, we would stop our other activities that day and remember the friend serving in that area. We would telephone parents, wives, girlfriends for news and close the conversation with best wishes. Then, inevitably, there were bodies brought home. The first funeral, we thought, was the hardest. A bright boy who always wanted to fly was the first among us to volunteer in the Army flight program. After a period of training and promotion to officer status, he went to Vietnam and was shot down. I remembered the days in grade school when we had shared a paper route and talked about our heroes, the Boston Celtics. Later, the Celtics did not seem very important any longer, nor did my—by then—postgraduate studies at university.

In the end, the worst of it was that the deaths and casualties stopped being desperately personal. The numbing effect of that was all the worse because it robbed us of our humanity. Later, as an unscathed survivor, I began teaching in the college at which I have remained. Older than my students by eight or ten years, I saw the pattern repeating of the dread that the scourge of war brought to their lives. The students who were themselves returning veterans—a few with physical scars, some with deeper, psychological scars—were a present reminder to the younger students that this matter of "the war" was not an academic abstraction. The protest and rallies grew in intensity until May 1970 at Kent State University. The war had "come home," and there were dead bodies of young people on campus, mirroring the dead in far-away jungles. It was very difficult in that time to carry on our normal work and live our normal lives. Now, a generation later, when students have no direct experience even of national service, and surely not war, they find it easier to get on with life, to make career and family choices, in short, to be energetic. They do not say, "We have never known war directly, so we are fortunate to be able to direct our energies elsewhere." Rather, they rarely think about it—and I do not fault them. But they go about their lives with a singular purposefulness that was not as possible a generation ago.

In the nineteenth century we note a sense of purposefulness and activity that both astonished people in the North Atlantic world and encouraged them to think that progress was being made. The "new men" we discussed in the previous chapter are explicable in these terms. When people believe that purposeful activity will have a beneficial result, they can give their minds and spirits to it. And, without a negative drain on them, these people believed they were bringing forth a new world. Moreover, there were signs of moral improvement that further encouraged people. Early in the century, slavery was abolished in the British Empire, and later it was abolished in two large slave-holding nations, the United States and Brazil. The serfs in Russia were also freed, and human life seemed to be valued in other ways as it had never been before. In sum, the people of the nineteenth century were not fools to believe in progress because, in some substantial way, they had experienced a century of relative peace.

The Nineteenth Century Was a Century of Scientific and Technological Progress

While relative peace may have provided a partial context for a belief in the reality of progress, the proof of it was in the signs of everyday life. The standard of life rose continuously throughout the nineteenth century, driven by an apparently uninterrupted advance of science and technology. We will return to developments in "pure" science in the next chapter, but for now it is enough to say that the growth of science is culturally important insofar as it fuels the fire of technology, i.e., the application of science to practical concerns. The explosion of new inventions and the businesses to put them to practical use was itself a substantial sign, for most people, of advancement and enough to validate the belief in progress.

Let us imagine, for example, a person born in 1820 whose father may have served in the Napoleonic wars. During the range of a normal lifetime, three score years and ten, the person would have witnessed an astonishing transformation of society: from horse-drawn carriages to the motorcar of Karl Benz; from rudimentary modes of transport on sea and land to the oceangoing vessels and railroads powered by coal; from subsistence agriculture to capitalist agriculture; from a countryside of small farmers to cities of teeming millions of industrial workers and artisans; from water supply and sewage practices that had not changed for centuries to the beginnings of pure water and sanitary conditions; from candles to interior lighting; from nonexistent or poor medical care to the professionalization of medicine. Other examples could be listed, but these are enough to suggest the transformation that had taken place in northern Europe and North America from, say, 1820 to 1890. Not for nothing did a person whose life had spanned those decades believe that society *was* moving forward and getting better.

Demographically, there was a remarkable change. The population of Europe more than doubled during the nineteenth century because of several developments: a markedly decreased infant mortality rate; better nutrition, sanitation, and medical care. This was true in the United States as well, where natural increase was coupled with a massive immigration, resulting in a growth of population, from about five million in 1800 to one hundred million in 1900. But, as well as overall growth, the stabilization of family size by the end of the century was equally remarkable. Again, in the American example, in 1810 there were 1,058 children under the age of five for every 1000 white women of childbearing age. According to James C. Mohr (1978), by 1890, it had dropped to 685 per 1000. In other words, whereas the average family at the beginning of the century had seven children, by the end of the century it had three or four. Contraception had become more widely known and practiced as the century progressed. It seems that abortion was a significant means of birth control. As the available figures seem to indicate, during the nineteenth century in America, at least one-fifth of all pregnancies were aborted, with Michigan having the national high at thirty-four percent.

While human motivations in population growth and, then, stabilization are difficult to discern, social scientists offer the following line of thought as hypothesis: It seems that the availability of regular medical care (the American Medical Association was founded in 1847) gave people an increasing sense of control over their own lives, and there was no longer the need to have ten children if you wanted five to survive till adulthood. As the population urbanized, there was no longer the economic benefit of a large family as in rural areas. There seems to have been some conscious choice about limiting the size of families. Part of that choice seems also to have involved the desire to maintain the higher standard of life that had accompanied the rise of wages throughout the century. Moreover, this higher standard of life brought goods and services, previously available only to the wealthy, into the hands of ordinary people. Pioneers in mass marketing, such as F. W. Woolworth, Aaron Montgomery Ward, Richard Sears, and Alvah Roebuck, brought consuming to a mass culture. Not for nothing did ordinary people believe that their lives were getting better.

Science and technology created new things and new situations for people who believed that these creations were harbingers of a new and better life. As the nineteenth century wore on, people in the cities and in the countryside, whether in Norway or in Minnesota, in Germany or Missouri, in Britain or in New York, sensed that the great things happening in industry, transportation, and medicine had something important to do with their lives. This, coupled with the relative absence of war, encouraged people to believe that the world could become a better place and that the Enlightenment hope for "happiness" might be realized.

The Nineteenth Century Was the American Century

America, as we observed in an earlier chapter, entered the European mind as a symbol of renewal and hope. The "idea of America" preceded the existence of the nation-state called the United States, and, ironically, Europeans held the United States up to the example of what "America" was supposed to be. If hope for the progress of humankind is one of the main themes of the nineteenth century, that hope was deeply involved with the ideal of America becoming a reality in the United States.

Deep in the consciousness of European thought and feeling, and even more primordial than the Christianity that was formative in Western culture, is the deep human desire to begin again. Who does not wish to have a second chance, to have known in youth what one knows in old age? Who does not want to enter into life again but to do it all better? In short, who does not want to be "born again"? For good or ill, America was that land of possibilities to which Europeans looked, and, in some substantial senses, America was to be the proving ground for the ideals of the Enlightenment. Further, what Europeans first hoped for, the Americans later insisted upon for themselves. Thomas Jefferson, ever the Enlightenment man, repeatedly called America "an experiment" in the possibilities for humankind. Later, Walt Whitman, the greatest poet of America, would assert that America was destined either to be the greatest success in the history of the world or its worst failure; nothing else would do. The experiment in America, the laboratory for humankind, would be proven true or false, for all time. America was more than a nation; it was to be the secular hope of the world.

When the Puritans came to Massachusetts in 1630, they were led from England by John Winthrop. Sighting land late in the afternoon, they decided to wait to come ashore until the following morning. Assembling the whole company of emigrants in the hold of the ship, the *Arabella*, Winthrop spoke to them of their purpose in coming to the "new world." He encouraged them to see themselves as more than a few emigrants but as players in a larger, moral drama. In his sermon, called "The Model of Christian Charity," he outlined certain ideas that were to be formative in the American character. As literary historian Sacvan Bercovich (1969) has suggested, this sermon, when repeated and referred to by later generations, helped to create "the puritan origins of the American self." The eyes of the world, said Winthrop, would be on New England because the Puritans were sent by God to do "an errand in the wilderness" that would be a boon or blight to later generations. They were to be "the city on the hill," the "light to the nations," in respect of whose shining all would one day turn and be converted. But this could only happen if they kept the covenant given them by God: If they kept faith, God would bless them, and future generations would emulate them; if they broke the covenant, it would be worse for them than for other nations and peoples because they would have rejected God directly.

It is not that one sermon made an ideology but that the "New England mind"—in all its aspects—reinforced an idea. In time, this became the American idea, that America was exceptional in the history of the world and that on its fate hung the fate of humankind. As historian Harry Stout (1986) has pointed out, the Puritan sermon is an important part of the American literary heritage, and the sermon goes a long way in establishing an American "sentiment." That sentiment has been sustained throughout American history, although, by the late nineteenth century, shorn of its explicitly Christian content. The Enlightenment ideal of human improvement supplanted the "Christian-covenantal" aspect, but the overriding ideology of American exceptionality was maintained.

As we have already seen in the chapter on the Renaissance, there is another European tradition to which the Protestant one is indissolubly linked: the humanist tradition. And right from the beginning of "the discoveries" it was hoped that the "new world" would produce a "new man." So, for the so-called Christian foundations of America to become secularized in the eighteenth and nineteenth century is, really, to have it return to its humanist origins. After all, Christians are not, in the first place, citizens of a nation-state but citizens of the kingdom—a kingdom that transcends all geographical, racial, and temporal boundaries. To have asserted that God had a special purpose in one nation—for America to be the "new Israel"—was always wrong because it blinded Christians in America from seeing, and acting upon, their essential solidarity with Christians in other nations. In biblical terms, the modern nation-state is never coincident with the kingdom. Israel of old was indeed a people chosen by God. But in the post-Resurrection era, "Israel" is the people of God wherever they are found, i.e., among all tribes, tongues, and peoples. Those who name the Name and follow Christ in life *are* God's people, and the Bible cannot be twisted to give that position to the people of a single nation-state.

The ideal of America as hope of the world nevertheless has continued throughout its history, both because Europeans wanted it so and because Americans themselves accepted this historic role. The great stake that Europeans had in American exceptionality was stated in a classic book by Alexis de Tocqueville (1805–59), a young French nobleman who visited America early in the nineteenth century. His book appeared in two volumes, the first in 1835, the second in 1840, and *Democracy in America* has been rightly hailed as the best book ever written about America. In a much-quoted passage in *Democracy,* Tocqueville sets forth an important point germane to the theme of this chapter.

It is not then merely to satisfy a legitimate curiosity that I have examined America; my wish has been to find there instruction by which we may ourselves profit ... I confess that, in America, I saw more than America; I sought there the image of democracy itself, with its inclinations, its character, its prejudices, and its passions, in order to learn what we have to fear or to hope from its progress.

Hope and fear are the most essential human emotions, and linking European hopes and fears to the success or failure of America is to establish its primary importance. If America succeeded, there was hope for humankind, but not if she did not. No modern nation has ever had to bear the weight of such idealism as this.

This belief in "America as hope" was my no means confined to the writing of intellectuals in Europe. It was agreed to by the millions of ordinary people who emigrated between 1815 and 1914. In that century, Europe gave up some sixty million people who migrated overseas. About two-thirds of them came to North

America (here Canadian readers must not take it as a national insult to realize that nineteenth-century emigrants regarded the United States and Canada alike as "America"). They emigrated for a wide variety of reasons—social, economic, religious, politicals—but, taken together, the motivation for their coming was to seek what they called "a better life" in the new world. It is, of course, a cliché to observe that America is a nation of immigrants. But it is true enough that, even among other immigrant nations (Australia, Brazil, Argentina, South Africa, etc.), the United States received more, and more ethnically diverse, people than any other nation.

There is no theme more deep in American consciousness than that of the transplanted person who comes to participate in the American experiment and who succeeds in the land of the free. Indeed, America has only a few symbols that represent her history, and primary among them is the larger-than-life statue, standing at the main door of the nation, proclaiming its most essential quality—liberty. The Statue of Liberty in New York is the icon of America, and for many immigrants and native-born citizens alike, it represents the essence of the national spirit, i.e., a free nation of diverse peoples, doing a new thing in the new world where they could sing a new song. For those with hearts to feel and with ears to hear—represented by Walt Whitman—one could "hear America singing" the lusty carols of vibrant democracy, among amber waves of grain and in alabaster cities.

Abraham Lincoln—the most essential American—picked up the theme of American hope in the context of the Civil War. The stakes in that war were high because at question was whether or not a nation dedicated to liberty could endure. And vastly more important than who won the Civil War was whether or not America succeeded, whether or not "the last best hope of earth" could be saved. America divided, and with slavery, was not America at all but a travesty of its ideals. This theme has been a constant one throughout American history, most recently echoed by Richard Neuhaus (1984), who suggested that, as much in the Cold War as in the Civil War, the survival of a free and liberal America was much more than a matter of national concern; it was a matter of world concern because (whether Americans like it or not) America is the "sign society" for the world.

To return to the theme of this chapter in conclusion, the American experiment was an important part of the widespread belief in the idea of progress. America, the land of possibilities, allowed men and women to dream of what had not yet been but might yet be. A much-quoted and much-remembered epigramatic rendering of American "possibility thinking" was often repeated in slightly altered form by the late Robert F. Kennedy: "You see things, and you say 'why?' But I dream things that never were, and I say 'why not?' " While these stirring words may well encapsulate the American attitude toward life, it is important to note that they were written by a European, George Bernard Shaw, in his play *Back to Methuselah* (1921). The opening scene takes place in the Garden of Eden, and the character to whom Shaw gives the question is no less than the serpent.

Finally, this book—premised on its Christian perspective—wants to avoid simple judgmentalism; but, at the same time, a critical assessment of the idea of progress needs to be given. We have clearly seen the Enlightenment roots of the idea of progress, and those are planted in different soil than the Christian tree of life, of which Christians are the branches. The premise, after all, of the Enlightenment was that humankind, of its own nature, was not essentially flawed but corrupted by unnatural institutions. Further, that premise allows a logic of liberty to develop wherein great hopes for humankind could be held out if people were liberated from unnatural bounds. Human progress was to be the result, and, as we have seen, there was a plausible basis for belief in progress. But—and this is vital—the idea of potential in the Promethean human when unbounded is an old one in Western culture. It first challenged the Christian church in the fourth century. A Celtic monk named Pelagius tried to reconcile the legitimate

Christian hope for the future with a Greco-Roman hope of human boundlessness. That view was condemned as a heresy. When we hear about the faith in human progress in the modern world, we who take a longer, Christian view of things cannot help but remember the Pelagian controversy and that Christians have had difficulty with it for some time. The fact that many Christian readers of this book in the United States will recognize a kind of "Pelagianism" in American ideology will, perhaps, have something to say to whether or not they can highly value assertions that America is a "Christian" nation.

Chapter 14

THE POST-MODERN MIND: THE
SHAKING OF FOUNDATIONS

Most historians of culture now agree that there was a "crisis of the spirit" beginning in the middle of the nineteenth century and continuing into the early years of the twentieth century. While the full dimensions of it are not yet fully clear to us, and a full treatment of it is beyond the purview of this book, we can discern something of the outlines of that crisis. There were intellectual movements in the arts, the sciences, and the then-new social sciences that caused considerable doubts to be raised about the entire worldview we have called the "modern mind." Let us recall that by modernity we mean, in this book, those patterns of thought and behavior that came out of the Enlightenment. While it is not exactly a precise term, *post-modernity* represents those patterns of thought and behavior that came after the Enlightenment had developed as a worldview. The new patterns are rebellions against the Enlightenment, even calling its main tenets into question.

Whatever the Enlightenment meant in its various forms, indisputably vital in its worldview is the primacy of reason. Paradoxically, the intellectual movements we will discuss here are, taken together, a revolt against "intellect." Or, put another way, what is called into question is the primacy of reason, which had been the hallmark of the modern mind. One of the deep currents in the twentieth century is *nonrationalism,* by which we do not mean *irrationalism* but a conscious rejection of the faculty of human reason as the highest state of human endeavor and the best avenue toward truth. This chapter will discuss the revolt against intellect because nonrationalism erodes the ground under the structures of modern thought and behavior. And when the tumultuous events of the twentieth century shook the modern structures, the foundations were not able to hold. As in the story of the house built upon sand, it stood firm until the storms came, but it could not withstand them. While the above is surely too simplistic an analogy, the Enlightenment worldview nevertheless was called deeply into question by romanticism, biology, and psychology. When the events of the twentieth century pounded modern ideology, it was inadequate to the test because of the erosion that had gone on during the spiritual crisis of what in Britain is called the late Victorian period and what in America is called the Gilded Age.

Romanticism: The Irrelevance of Intellect

Having said that the new currents of nonrational thinking emerged in the Victorian period, we must immediately qualify that by noting that the so-called Romantic movement had origins earlier in the late eighteenth and early nineteenth centuries. While it is true that Romanticism's full impact would not be felt until the advent of a mass reading public after about 1870, it is also true that Romanticism's rebellion against intellect was a long time germinating. The German composer Franz Schubert, early in the nineteenth century, called the Enlightenment an "ugly skeleton without flesh or blood." What he meant was that the whole universe of the *philosophes* was too easily reducible to mathematical precision. Indeed, as we have already seen, the early *philosophes,* and their scientific forebears, did try to speak in Nature's mathematical language. Moreover, many of the *philosophes* were either indifferent to, or opposed to, religion. On the other hand, romantics saw religion and faith as part of human existence, and rather than trying to deny the role of the spirit in humanity, they sought to exalt it.

What we mean by "romance" in everyday life is something like the historic movement called Romanticism. The stuff of romance—love, sorrow, joy, pity, passion—are the stuff of the Romantic style. The most important aspect of "real" life to Romanticists is that it cannot be explained in ordinary words and analytical ways. When I was a student, a popular song said, "There were bells on the hills, but I never heard them ringing . . . till there was you." Now, Thomas Jefferson, who believed in self-evident truths, would have heard the bells in the first place. Ringing bells are self-evident. But if they can only be heard in the presence of a special person, something strange has happened to the nature of reality. To have eyes to see what others do not or cannot see, to have ears to hear what others do not or cannot hear, to have a heart to feel what others do not or cannot feel is to open oneself to a special reality.

What was at stake, at the deepest level, in the Romantic movement was the alienation that many creative artists felt in a world of mechanization. As much in the level of theory (the rationalism of the Enlightenment) as in practice (the increasingly ordered technological world of industrial society), creative persons felt stifled by the very mechanistic conception of the universe that was supposed to explain all things. Whether in Victor Hugo's *Hunchback of Notre Dame* or Walter Scott's *Ivanhoe* or Johann Wolfgang von Goethe's *Faust,* writers brought out ahistorical themes in far-away and unlikely scenes and places. Whether in Hector Berlioz's operas, in Ludwig von Beethoven's symphonies, in Frederic Chopin's piano concertos, or in Nicolo Paganini's violin concertos, composers aimed at the emotion, not the mind, to evoke a sense of release in the listener that does not come from, say, the technically more "mathematical" music of Franz Joseph Haydn. Whether in the paintings of the idyllic English countryside by John Constable or those of the noble deeds of the Middle Ages by Eugene Delacroix or the swirling seascapes of J.M.W. Turner, Romantic painting transported the viewer to other places and times beyond the more mundane aspects of urban-industrial reality to a place where "consciousness" was said to be higher. Whether in the poetry of William Wordsworth or of Ralph Waldo Emerson, readers were encouraged to regard the grand and noble in nature to be "real."

The religious revivals of the nineteenth century must be seen as part of the romantic movement. It is well for us to recall an evolution of the locus of religious authority: from medieval Catholicism's emphasis on the authority of the church; to the Lutheran and Calvinist emphasis on the (relatively rationally known) authority of the Bible; to the deists, who saw religious authority in Newtonian first and last causes; to the Romantic religious leaders who saw religious authority founded in human emotions. Romanticist religion ought to be seen as part of a larger cultural rebellion against the "cold" and "formal" aspects of culture of the eighteenth and nineteenth centuries. Many examples could be given from Britain, the most advanced technological nation: Important in this regard is the work of John Henry Newman and the revival of "Anglo-Catholicism," but the best example is Methodism.

Methodism began in the mid-eighteenth century, under the leadership of John Wesley (1703–91), as a rejection of deist and rationalist religion in the Church of England. After studying at Oxford, Wesley went out to America as a missionary in 1735. During a violent storm at sea, when Wesley thought his life about to be lost, he was much impressed with the spiritual serenity of some of his fellow-passengers, Moravians from Germany. After an unfulfilling three years in America, Wesley returned to Britain, and, in addition to regularly worshiping in his own church—the Church of England—he worshiped with the Moravian Brethren in London. In 1739 he experienced a conversion unlike anything he had known before. At that time, he uttered words that have become sacred to Methodists the world over: "My heart felt strangely warmed."

Wesley was not welcome in the Anglican church, which had little appreciation for his theme of conversion that resulted in piety. After 1739 he began to preach in open-air services mainly in the west of England. He preached to great effect among the people of what were called the "lower orders" of society, and this preaching was complemented by the stirring singing by the people of the songs composed by Wesley and his brother, Charles. At first not wanting to create a new church, they formed Methodist societies in the parishes of the Church of England. After a few years, the societies became churches in their own right, and Methodism became strong enough to send missionaries to the North American colonies. The greatest success of Methodism was to come in the United States. First mostly in Virginia but then throughout the country, the Methodists were highly motivated with their conversionist ideology and better organized through a system of circuit riders to adapt to the moving frontier. Statistically insignificant at the time of the nation's birth, they were the largest denomination by the time of the Civil War. As historian William W. Sweet (1950) has written:

The doctrine preached by the Methodist circuit riders was also well adapted to meet the hearty acceptance of the frontiersmen. It was a gospel of free will and free grace, as opposed to the doctrines of limited grace and predestination preached by the Calvinistic Presbyterians, or even the milder Calvinist Baptists. The frontier Methodist preachers brought home to pioneers the fact that they were the masters of their own destiny, an emphasis which fitted in exactly with the new democracy rising in the west, for both emphasized the actual equality among all men.

For the purposes of this chapter, it is important to note that Methodism, as a religious movement, has interesting convergences with the Romantic movement. It is not so much anti-intellectual as not stressing the intellect. The appeal of the gospel in Methodist revivalistic conversion was to the heart. Revivalism— later to be called evangelicalism—could be anti-intellectual, but it need not necessarily be so, as early Methodism showed. Indeed, this brings up an important point, vital to the argument of this book. Christians were—and are—committed to a "whole" view of life. Any account of the human past and future that does not take into consideration the spirit as well as the mind is incomplete. Methodism's historic importance for the history of civilization (other than its obvious religious influence) was that it offered a counterpossibility to the main story of the intellectual conquest of Enlightenment rationalism. Methodism, in short, attempted a balance between "heart" and "mind." When, in later chapters we see the bankruptcy of Enlightenment ideas, it will be to the sorts of concerns raised so successfully by Methodists and others that we shall turn. Yet, as a historical matter, Methodism's main appeal and the reasons for its initial success cannot be seen apart from other aspects of culture in which faith in "mind" was being devalued in respect of a higher valuation upon the emotive and nonrational.

Darwin and Biology: The Irrelevance of Purpose

As philosopher-historian Morton White (1957) has suggested, the intellectual currents of the late nineteenth century were "a revolt against formalism." In the terms of this book, *formalism* can be regarded as the Newtonian world picture, i.e., that the physical world was rational, mechanical, and dependable. In this regard, the work of Charles Darwin (1809–82) in biology is both affirmation of mechanism and revision of a formalist emphasis on reason. His *Origin of Species* (1859) is a monument of modern science, and, as to continuing importance, the equal of Newton's *Mathematical Principles,* published nearly two hundred years earlier. It is of vital importance to recall that the earlier work in modern science had mostly gone on in the physical sciences, and only later is there important work in the life sciences.

Darwin brought a mechanistic interpretation of life to the world of living things. But rather than bringing a grand theory to reality, Darwin began with a very practical set of questions that had come out of his direct observation of the variety of animal life in the world. He circumnavigated the earth on a scientific expedition, and it struck him that in certain parts of the world the species of living things varied enormously. His life's work was an attempt to explain this variety. After long years of studying plants and animals he published his book in 1859, and, the conversation about the nature of life would never be the same again. While evolution as a theory existed before Darwin's book, he gave clearest expression about how it might have happened.

In brief, Darwin's explanation runs thus: Variations among species were hereditary, and the offspring of living things developed, over time, those variations that would allow them to survive in their environment. Since all living things must, in the first instance, seek constant physical nurture and since there is always pressure on all ecosystems to support the life in it, life was, for these species, "a struggle for existence." Embedded in natural laws were certain mechanisms that "selected" certain species to survive in the struggle. So, certain species were "unsuited" to survive and would disappear, while others were able to adapt, especially to changing environments, and thus they were "the fittest" in terms of survival. While all the "links" in the chain of development were not fully clear, it seemed obvious that "life" (taken in its entirety) had moved through an evolutionary process from the simple to the complex and that humankind was the pinnacle of evolutionary advance.

Reactions to Darwin were—and continue to be—mixed, and the debate over evolution is a perennial one in the conflict between science and religion. Were it only *Origin of Species*, Darwin's work might not have received the reaction it did. But twelve years later he published *The Descent of Man* (1871), and in it he continued his thinking, this time specifically about humans as a species of higher animals. Moreover, it was not merely the physical nature of humans—but moral as well—that were adaptive responses to the struggle for survival. To this, Christians (and other believers in the existence of a human "soul" independent of natural life) reacted in horror. To them, Darwin's views represented the ultimate devaluation of humankind and ones that were totally incompatible with a view of humankind as created in the image of God. For humans to have had "brute" origins (in the language of nineteenth century anti-Darwinians) was to question the very nature of both God and humankind.

It should be noted that some Christians *have* found a way to reconcile Darwinian biology and Christian belief. They accept, with Darwin (who wrote in *The Descent of Man*), that "the grounds upon which this conclusion rests will never be shaken. They are facts which cannot be disputed." What is possible for some Christians is the stance taken previously by Newton, i.e., that the *manner* of human life's creation most probably is that which the scientists have discovered, but the *cause* of the process is one of divine agency. The God of first and last causes is plausibly invoked in this connection. Even the Christian Bible says that God used "dust" to create humankind. Why, they ask, could not the biblical metaphor of "dust" be explained in more precise scientific terms? If God himself was not embarrassed to create out of dust living things that looked like him, why should those very creatures be outraged when others specify what "dust" was and the process by which "dust" became "human"? As historian James Moore (1979) has shown in his important book *The Post-Darwinian Controversies,* it was generally those Christians who had a high view of God's sovereignty that had an easier time accepting Darwin. In short, the God of timelessness and infinite power and grace "moved in mysterious ways his wonders to perform." The humans who look like him can also (though in a limited way) think like him, and they have discovered (been allowed to discover?) the manner in which God set his creation in motion and allowed it to develop from simple to complex. Whether or not this moderate stance of reconciliation between Darwinian biology and Christian belief is an adequate one, readers of this book will decide for themselves. As is well known, the debate between "Darwinian evolution" and "special creation" continues to our own time.

It should be stressed that what is of greatest importance in Darwin to theists is not "evolution" versus "special creation." Of much deeper significance is the assumption that the world, in reality, was in constant flux. This is an important revision of the Newtonian world picture that believed that there were fixed, immutable, and changeless laws that worked their implacable will in nature. Darwin also believed in the unstoppable working of natural laws, but they were *dynamic* not *static*. The physical and organic nature of things was subject to constant change; thus, one natural law that worked at one place and time might not work at another. The key to success, after all, was adaptability because the circumstances of life changed over time. Since humankind is not immune to natural life, humans must develop skills that will allow them to adapt as life grows increasingly complex. This has important implications for religion and society because one infers from this line of thought that there can be no fixed propositions that are true "yesterday, today, and tomorrow." Rather, there is only that skill of adaptation that will get us to tomorrow, and then a new dynamic takes over that requires a new adaptation from us. In this regard, student readers of this book will want to inquire of their education professors about the implications of Darwin for education (John Dewey) and of their politics professors about the implications of Darwin for law (Oliver Wendell Holmes).

The world is, according to the Darwinian way of thinking, in a process of continued dynamic change. This change, moreover, cannot be stopped but only accommodated, and this accommodation is *never*, in the first instance, an intellectual one. In short, human purposing cannot alter the immutable process of dynamic change. So, although the implications of Darwinian biology are, on one level, affirming of a naturalistic, even mechanistic, view of the world, they undermine substantively the rationalist contention that "mind" acts independently of, and purposefully on, natural phenomena. For Christians this would seem to be a more far-reaching and unacceptable implication of Darwinism than that of the conflict between evolution and special creation.

Positivism and Psychology: The Irrelevance of Rationality

The term *evolution* is linked indissolubly with the name of Charles Darwin. But a contemporary of his, Auguste Comte, developed a sort of theory of evolution that may be of even greater impact than that of Darwin. Comte is the father of *positivism*, and, while most academic intellectuals would not identify themselves specifically as *positivists*, the matrix of contemporary intellectual developments is positivistic. For this book, seeking a Christian perspective on the history of Western civilization, positivism presents the greatest single challenge to the book's integrity. Indeed, a positivist would think the attempt of this book is impossible if one were intellectually honest. Since positivism is such a challenge to Christian thinking, we must look into it and its consequences.

Such was the prestige of science in the nineteenth and twentieth centuries that scientists could plausibly say that humans could, indeed should, "come of age" by putting aside immature views of life and by seeing life in a mature (i.e., scientific) way. The most important spokesperson for this view was the originator, Auguste Comte. In *The Positive Philosophy* (1842), Comte argued that humans have tried to understand reality in three ways. The first way was theological or supernatural, in which humankind invented a "spiritual dimension" (gods, spirits, forces, etc.) to cover their lack of a "real" understanding of things. Second, humans progressed to the philosophical or metaphysical dimension, in which investigators replaced "gods" with "principles" or abstract ideas about the nature of the world. But since, in Comte's view, knowledge accumulates, it was time for humankind to realize how far they had come and to accept that they could put aside "gods" and "abstractions" and draw "positive" conclusions about life, drawn from a direct, empirical examination of life itself.

Although *sociology* was not yet a word when Comte wrote, he is often regarded as "the father of the social sciences" because he was among the first to believe that a positive knowledge of humankind was possible if the scientific methods of empirical inquiry were thoroughly used in the human realm. This is a frankly materialistic philosophy (that is, antispiritual), and the degree to which this method of knowing dominates academic discourse is, perhaps, the reason for the existence of this book. Let us be clear what we mean by saying this: It is not that contemporary scholarship believes religion unworthy of academic investigation (there are societies to study churches and the phenomenology of religion), but what *is* unworthy and intellectually dishonest, positivists say, is to try to study anything from a religious perspective. If historians and other analysts of society and culture learned of this book, they might well say, "What? A 'Christian perspective' on the history of civilization? Surely you can't be serious. That is an antiquated way of looking at things. This is the twentieth century, after all." Even if these sentiments are put a bit more subtly, the degree to which they are said at all is the degree to which Auguste Comte lives. Sociology and psychology were meant not merely to be new academic disciplines and to take their place among the others; they were meant to supplant the older forms of discourse. To Comte, theology was surely outmoded, and, even within the sciences, he made a list in terms of the order of maturity—and of the potential—of the sciences: It goes from mathematics and astronomy through physics and chemistry, finally to the "newer" sciences of biology and psychology, and ultimately to sociology.

Psychology was to be for the person what sociology would be for society, the ultimate and "positive" science of humankind. Psychology is the more important because of the widespread acceptance of the work of the Austrian medical doctor Sigmund Freud (1856–1939). The irony of Freud's work in terms of Comte's hope was that Comte's vision for the future of humankind was based on empirical science, but Freud's scientific work was—most vitally—to call into question that staple commodity of scientific thought, human rationality.

As a medical doctor, Freud approached human disorders in a classic rationalist fashion, believing that good and sufficient reasons underlay the disorders that came to the surface. Still in a very scientific-rationalist mode in his early career, he sought new scientific explanations for illnesses for which ordinary physical causes could not be determined. He became interested in hypnosis and studied the technique in the 1880s under Jean-Martin Charcot in Paris.

Moving beyond hypnosis in the 1890s, he changed his methods so as to allow his patients to talk uninhibitedly about themselves. In the free associations of such disclosures he noted the recurrence of sexuality in the life histories that led to disorder. In an age that was loath to talk about private matters openly, Freud was a pioneer in establishing the place of sexuality in both order and disorder in human personality.

Even more important than his work on hypnosis and sexuality was Freud's organizational scheme of the interior of the human mind. In his most important work, *The Interpretation of Dreams* (1900), and in other studies, he gave the terms that are the stock-in-trade of our discussion of personality—*id, ego,* and *superego.* In general the id was our basic "animal" desire for physical and sexual gratification; the superego was the moral and expectational realm, largely imposed by society; the ego was the mediator between the two. In dreams, the id might hold unrestrained sway, while in the daily life of wakeful hours, the ego had to "repress" the "drives" of the id to accommodate the demands of the superego.

Insofar as Freud's work was known—and it was widespread—it caused a considerable revision of what people thought human personality was. His findings were, in a substantial sense, contrary to his own predispositions. He was, in many ways, a nineteenth-century scientific rationalist and believed that religion was an illusion. But his research into human personality convinced him how difficult it was for people to act "humanly" and "rationally." It is a dishonest caricature of Freud to say that he advocated a liberation of the id over the superego, or, in colloquial language, "to let it all hang out." Because, while Freud wanted and hoped that humans would act rationally—personally and in society—he knew that humans were not essentially rational beings. A metaphor associated with Freud, like Newton's clock, is that of an iceberg. The tip of the iceberg is what we see above the surface of the ocean. But what propels it are the currents deep beneath the surface, unseen and powerful. Human personality is like that, insofar as we come to know that what really makes us "tick" are mainly not the rational acts of our waking hours but rather the sum of the experiences of our total lives, most of which lay unseen and little understood in the recesses of our subconscious. The impact of Freudian insights on the humanities and social sciences has been broad and deep, and his work represents a major benchmark in the history of Western thought.

In conclusion, let us see where we have come. In the late nineteenth and early twentieth centuries there were important movements in the arts, the sciences, and the social sciences that, taken together, represent a significant revision of modern thought. What is known as postmodernity arose out of a rejection of the Enlightenment's key point, the primacy of rationality. This, paradoxically, was an "intellectual revolt against intellect," or an assertion of the nonrational over against the rational. This was important for two reasons: These postmodern patterns of thought had tremendous impact on twentieth century believing and behaving; and an important result was that, in emphasizing the *nonrational,* the new patterns of thought were a fundamental erosion of the ground on which modern institutions had been built. When the events of the twentieth century thundered the storms of disorder, it was not clear that a worldview—or the society it created—based on order, rationality, and progressive hope could endure.

Chapter 15

THE TWENTIETH CENTURY:
THE AGE OF ANXIETY

We do not yet have a commonly accepted name for the twentieth century, like the Renaissance or the Enlightenment. The changes in our time, however, are no less momentous than in the earlier periods of change. Moreover, scholars do not yet fully agree upon the cultural changes in the twentieth century nor on the meaning of the changes.

Anxiety is a plausible name for our time because it suggests a mood that, though hard to define exactly, is deeply felt throughout Western culture. It is a mood whose internal character can be described in a developmental way, roughly thus: Once upon a time, when we were children, we could believe in certain things. But when we became adults we had to give up our fantasies and live in the world of realities. Usually events crash through our make-believe worlds and cause us to rearrange our sense of realities. In Western civilization, what people have had to give up—what events have called into question—is the belief in progress, especially human betterment, that the Enlightenment had promised.

In this chapter we will look at some of the main events of the twentieth century. But, let us keep in mind the purpose of this book—to think Christianly about Western civilization. And with this in mind, we will not so much tell the story of the events of our time but seek the meaning of them in terms of a Christian worldview.

The "Great War" and the "End of Innocence"

Innocence in the sense we mean it here has to do with certain assumptions with which most people began the twentieth century. They were beliefs in human potential and in Western society's ability to progress to ever-ascending heights. As we have seen in a previous chapter, there *was* a plausible basis for belief in progress and people *were not* fools to believe that society was improving. So, when we say that such beliefs were shattered, we are not making a moral judgment about naive people, nor are we calling them foolish to have believed. Yet—and this is the hard point to accept—we see the people of the first half of the twentieth century trying to reconcile their hopes to bitter events, and we join them in throwing up our hands. We, too, would have liked to believe in the Enlightenment's promises. We, too, would have preferred to live in a world at peace among neighboring states whose peoples look forward to a steadily improving standard of life. We, too, would have preferred to avoid the pain of twentieth-century events. But, events overtook hopes, realities crowded in, and, for those who survived at all, life had to be revalued.

In the abandoning of innocence—whether personal or societal—the first major episode is, in retrospect a "great" event. The first occasion of a breach of trust is more important than subsequent events, even if those events are more severe. For a child or young person, the first time she learns that all of life cannot be responded to in childlike trust will be remembered later—and in the recollection of subsequent events—as the beginning of the end of an era. The "Great War" of 1914–18 was such an event. The fact that we have had enough wars in this century to number them does not take away from the fact that what we call the First World War was, in fact, the great war of this century. It was a great war not only because it shared the horrific qualities of all wars but because wars of its scale and duration were not supposed to happen in the twentieth century when humankind was supposed to be moving forward.

The facts of the Great War can be told easily enough, but assessing its meaning is the more difficult task that we attempt here. On the level of fact, Europe and Eurasia were divided up into powerful and confident nation-states (Britain, France) and four empires (Germany, Austria-Hungary, Russia, and Ottoman Turkey). Through a series of alliances these powers were arrayed against each other, and all possessed an arsenal of modern weapons never before used in warfare. It only remained for an event—the assassination of the heir to the throne of Austria-Hungary in June 1914—to spark the cataclysm that, in its severity, was to alter realities and perceptions in Western culture. At war's end four years later, the four empires named above had collapsed, communism had its first national victory, and even the "victorious" nations, Britain and France, had paid so fearful a price for victory that it was difficult to distinguish the victors from the vanquished.

When the guns were first fired in August 1914, they were different sorts of guns than those used when Europe had had its last general war, ending in 1815. The same technology that had transformed industry and agriculture throughout Europe had also transformed warfare. In 1815, rifles had been single-shot weapons; by 1914, there were small weapons that could fire continuously. In 1815, combatants rode on horseback; by 1916, the internal combustion engine could move armor-plated vehicles and send flying machines of destruction into the sky. In 1815, the only fouling of the air came from the odors of fear and death; by 1914, weapons of chemical destruction were released into the air. In 1815, the troops marched gloriously into battle en masse; by 1914, they learned there was no glory in the trenches on either side of the many parcels of "no man's land" for which they contended.

No one—in any nation or empire—was prepared for the realities of modern warfare, neither what it would be like nor what it would cost. In human terms, no one was prepared to believe in 1914 that the war would take the lives of nine million combatants, on land, sea, and in the air. Nor could the twenty million serious injuries to combatants be foreseen, nor especially the additional twenty million with some long-term psychological damage. Even further stretching our ability, we now know—but no one knew then—that in a total war there is little distinction between combatants and noncombatants. In this regard it is significant to recall that, because of blockades and bombings, civilian life, especially in central and eastern Europe, was disrupted. Many civilians were weakened by malnutrition and could offer little resistance to the epidemic of influenza that swept Europe in the last year of the war, leaving an estimated twenty million dead in its wake.

Such numbers overwhelm us as we try to understand what they could have meant. How shall we, several generations later, assess the meaning of approximately five million women whose husbands were killed in battle-related deaths? How shall we understand the meaning of approximately nine million children whose fathers did not return from the war? Our minds are numbed—even repelled—as we try to bring the large numbers down to a manageable scale. For example, here are some figures for the Battle of the Somme, which raged from July through November 1916: In the first day of the battle, on the British side alone, there were 57,470 casualties, of whom about 20,000 died. Yet, fresh troops—more often than not inexperienced, frightened young men—were poured into the battle. All through a relentless July, August, September, October, and November, British casualties averaged about 4,000 *per day.*

People on the home front were not fully aware of what was going on at the battlefront. The propaganda machines on both sides sanitized the news. Civilians sent off their young people with remarkable enthusiasm, at least at first, and their governments daily informed them of the glorious victories on behalf of "the fatherland" or "crown and country." But, by the end of 1916, the propaganda of neither side could conceal the realities.

The realities of war—always brutal—might have been understood and accepted had the costs been offset by some tangible gains or explained in terms of actual losses. But there seemed to be neither gain nor loss for the two sides paying the ferocious price. This apparent meaninglessness can be best illustrated by a few glimpses at the longest battle of the First World War. The battle around Verdun, France, raged most of 1916—in fact, from late February to early December. The cost was enormous, but neither side won nor lost. The fighting raged from the snows of winter, through the heat of summer, and back to winter again, but there had been little apparent gain for either side. A few of the French survivors described the scene as a living hell. They were pinned down by artillery in open fields that no longer sustained grass, flower, or tree. Death rained down on them from unseen quarters, night and day, with the fear of death equalled by the anxiety of waiting for the whine of an incoming shell. The fields turned to mud in the spring rains. New French troops came in daily down a road called *Voie Sacree* (the Sacred Road). In the spring the Germans mounted a major offensive, but the French held the high ground along what the soldiers called *Mort-Hommes* (Dead Man's Ridge). Some of the units lost nearly nine-tenths of their men, yet more came down the Sacred Road. In December, the Germans withdrew from the area. The death count was enormous at Verdun, nearly 1,400,000, or about 700,000 on each side.

It was not an easy thing to deal with 700,000 corpses. The Germans had the logistical ability to take their dead with them. But the French decided, since Verdun was on French soil, to bury their dead on the scene. When I visited Verdun fifty years after the battle, grass grew again on the fields and trees once again bloomed in celebration of life. But what remained was the overpowering sight of the largest cemetery in Europe. Standing in front of the modest museum built to commemorate the battle, I looked out over the gentle countryside and saw crosses row on row, marking the final resting places of a generation of French youth. I could not see where they ended because the horizon came up first. The figure of 700,000 dead is too large to take in. But the crosses, each with a name, were not. Walking among them leaves the visitor with a sense of the enormity of what went on in that place. Each of the dead had family members who probably have come to pay their respects. There seems to be no way to compass about in the mind the meaning of so much suffering and grief.

And, in any case, to what end all this sorrow and pity? For what purpose had so many given their lives? Heads of state and political leaders intoned the verities of national honor, but to the mass of ordinary people who had paid the price, the pious phrases were not nearly good enough. They had gone to war with certainties, but they came home with doubts. In attempting to discover the meaning of the Great War we do well to turn to contemporary poets.

The most celebrated and eloquent of the war poets was Wilfred Owen, who was killed in action in 1918, just before the armistice. He had seen some of the worst action on the British side, and his poem printed below is thought to be a classic. The title of the poem, "Dulce et Decorum Est," can be translated, "It is sweet and proper to die for one's country."

Bent double, like old beggars under sacks,
Knock-kneed, coughing like hags, we cursed through sludge,
Till on the haunting flares we turned our backs
And towards our distant rest began to trudge.
Men marched asleep. Many had lost their boots
But limped on, blood-shod. All went lame, all blind;
Drunk with fatigue; deaf even to the hoots
Of tired, outstripped Five-Nines that dropped behind.

Gas! Gas! Quick boys!—An Ecstasy of fumbling,
Fitting the clumsy helmets just in time;
But someone still was yelling out and stumbling
And flound'ring like a man in fire or lime . . .
Dim, through the misty panes and thick green light,
As under a green sea, I saw him drowning.

In all my dreams, before my helpless sight,
He plunges at me, guttering, choking, drowning.

If in some smothering dreams you too could pace
Behind the wagon that we flung him in,

And watch the white eyes writhing in his face,
His hanging face, like a devil's sick of sin;
If you could hear, at every jolt, the blood
Come gargling from the froth-corrupted lungs,
Obscene as cancer, bitter as the cud
Of vile, incurable sores on innocent tongues,
My friend, you would not tell with such high zest
To such children ardent for some desperate glory,
The old Lie: Dulce et decorum est
Pro patria mori.

A similar theme was taken up by the Irish poet, William Butler Yeats, in "The Second Coming" (1920):

Turning and turning in the widening gyre
The falcon cannot hear the falconer:
Things fall apart; the centre cannot hold;
Mere anarchy is loosed upon the world,
The blood-dimmed tide is loosed, and everywhere
The ceremony of innocence is drowned;
The best lack all conviction, while the worst
Are full of passionate intensity.

The Russian Revolution and the "Spectre of Communism"

The Bolshevik revolution in Russia was the second major blow to Western optimism. In fact, it cannot be seen apart from the growing Russian disenchantment with the policies of the tsarist government, especially in the Great War. The destabilization of Russia had come about because of the war policies of the tsars and because of a general war-weariness among both army and people. This brought to a head the general discontent that had been brewing in Russia for the previous half-century. From 1860 to the eve of the First World War, the population of European Russia doubled, causing great agrarian discontent. Also, the tsars, Alexander III and Nicholas II, were pushing a policy of industrial modernization, organized under the leadership of Count Sergei Witte

(1849–1915). The industrial revolution in Russia was as brutal—for the workers of the initial generation—as it had been in England or America. It is in the context of deep discontent in both the cities and the countryside that the tsarist war policies were increasingly unacceptable. They were unacceptable to worker and peasant alike because they believed—rightly, as we now know—that the motivation for tsarist war policies was one of trying to rally national unity by pointing public attention away from domestic discontent. The Russian wars against Japan and later against the Central Powers (Germany and Austria-Hungary) were equally disastrous. Almost in a replay of Paris in 1789, when Louis XVI would not really cooperate with the Estates General, Nicholas II would not really deal with the Duma, the Parliament, after 1906. The Russian Revolution was vitally linked to the First World War.

The leading Russian Marxist in exile was Georgi Plekhanov (1857–1918), who lived and wrote in Switzerland. His main disciple, and leader of the Russian Revolution, was Vladimir Illich Ulyanov (1870–1924), known to the world by the name he later assumed, Lenin. In Lenin's revolutionary life and thought, we see played out some of the contradictions we noted earlier in Rousseau's thought: of believing in democracy, yet insisting so fervently on obeying the "general will" of the people that it was necessary to force people to be free. Lenin argued for no compromise with the tsarist government. Lenin's views split the Marxist Social Democratic Party into two factions: one advocating evolutionary change (Mensheviks) and one advocating revolution (his own Bolsheviks).

Lenin, however, had no effect on the revolution that began in March 1917. The tsarist government simply was unable to rule and collapsed under the strains of domestic discontent and a lost war. In the chaos of early 1917, there were several parties contending for leadership. The balance was tipped toward the Bolsheviks through actions from an unexpected and ironic direction. Imperial Germany, still at war with Russia, sought to extend the chaos in Russia in order to end Russian participation in the war.

To this end, they brought Lenin to Russia from his exile in Switzerland. He and his friend Leon Trotsky (1877–1940) organized a *coup d'état* and, after one failure and flight to Finland, they were successful in November 1917. Even Leninist Bolsheviks later admitted that they never really dreamed that they would be so successful so soon. The Bolshevik government immediately ended Russian participation in the war, slowly consolidated its position at home, and emerged victorious in a civil war that lasted until 1921.

The "real" Russian Revolution and the cause of Bolshevik rule in Soviet Russia need not concern us here. The point of our concerns—the hammer blows that the events of the twentieth century gave to Western optimism—must turn to the "idea" of the Russian Revolution. As people in the West assessed its meaning, several points seemed to emerge. The Bolsheviks apparently had made a successful revolution without widespread popular support. The key to understanding the Russian Revolution seemed to be that context was everything. If an industrial society was fundamentally destabilized and authority up for grabs, a dedicated, highly disciplined force *could* seize power. People in the West began to pay close attention to the writings of V. I. Lenin.

Lenin had written as early as 1902 in *What Is To Be Done?* that the revolutionary cause should not be led by politicians within the destabilizing society but by professional revolutionaries in a secret and disciplined group that he named "the revolutionary vanguard." Moreover, Lenin believed that popular revolution might sweep Europe, even America, in the aftermath of the Great War. So, the work of "the vanguard" in other nations would be to replicate the conditions necessary for revolution: to destabilize "legitimate" governments. While focusing their energies on Russia, the Bolsheviks never lost sight of their worldwide aspirations, and they encouraged both the development of a Marxist-Leninist "vanguard" in all nations and the fomenting of civil disorder anywhere and everywhere. Many disillusioned intellectuals in Western countries were strongly attracted to Soviet communism, especially in the 1920s and 1930s.

One person in the West who took special notice of the Marxist-Leninist worldwide strategy was J. Edgar Hoover. In the 1920s he was a bureaucrat in the American Immigration and Naturalization Service. He believed, like most Americans, that if revolution came to his country it would come as a foreign menace. No native-born American, Hoover argued, would rise up against his native land. So, when the Federal Bureau of Investigation (FBI) was formed in 1924, with Hoover at its head, one of its main concerns was the importation of foreign radicalism. While the FBI did work against organized crime on the domestic scene, its main concern was with any and all elements that might destabilize America, thus creating conditions that might make revolution possible.

In this light we can see the long shadow of Lenin. J. Edgar Hoover and his FBI lived in a world that was the mirror image of Lenin's. Hoover feared that well-meaning elements in American society—in advocating reforms and questioning authority—might lead to a destabilization that, in turn, would give an opportunity to a Communist "vanguard." So, Hoover's actions on behalf of law and order over his half century at the FBI reflected not mere authoritarianism but a desire not to "play into the hands of the Communists." Thus, the civil rights movement as much as organized crime, liberal academics as much as drug pushers, social reformers as much as social deviants were subject to Hoover's scrutiny because he feared that, however unwittingly, they might make America ripe for a *coup*. Never mind that the Communist party was small; to Hoover, the context in which it worked meant everything.

The meaning of the Russian Revolution's mirror image can also be seen in the foreign policy of the West, especially the United States, since 1945. While Germany and Japan had been the prime focus of Western concern in the first half of the century, the Soviet Union has occupied that position in the second half. A recent president of the United States even referred to the Soviet Union as the center of an "evil empire" bent on expansion and world domination. Indeed, the major premise of American foreign policy, and that of the NATO alliance, has been the containment of international communism.

The purpose of mentioning all this in the present book is to see again that in the twentieth century themes of fear and anxiety are present that were not present in the nineteenth century. When Karl Marx wrote, in the beginning sentence of *The Communist Manifesto* (1848), "A spectre is haunting Europe—the spectre of communism," he wrote better than he knew or could have imagined. He could not have dreamed that a century later all the democracies of Europe and North America would indeed be haunted by the spectre of communism, centered on Russia. Most people in the West, or at least their governments, continue this day to believe that the world is a dangerous and hostile place for humankind, in large part because of what the Russians do or are capable of doing. The world is not a safe place to live, they say, because the red flag flies in Moscow, and, if we were not careful, would fly elsewhere soon enough.

The Second World War: Auschwitz and Hiroshima

Confidence and optimism in the West, already reeling from the pummeling of the Great War and the spectre of the Russian Revolution, received another blow—potentially fatal—in the form of the Second World War (1939–45). Most of the same parties who fought in the Great War fought in its reprise a generation later. For purposes of this chapter we need not discuss the battles of the war (titanic though some of them were) nor the "hellishness" of modern warfare (monstrous though it was for combatants). We might point out, however, that there was an increasing lack of distinction between combatants and noncombatants. Our concern here will focus on two places, one in Germany, the other in Japan. The events that happened in Auschwitz and Hiroshima illumine more graphically than we sometimes wish the chilling internal story of the second general war in this century.

Anti-Semitism has a long and ignoble history in Europe and elsewhere. Although Christians always have accepted the Jewish origins of their faith, an affinity for the Israelites of old did not always translate into positive feelings about Jews in one's own village or nation. At many times throughout Western history, Jews were denounced in religious terms, i.e., as those who had rejected Jesus. In a society, especially in the Middle Ages, that thought itself to be "Christendom," there was little place for those who rejected Christ once and rejected him still. Jews were systematically relegated to the margins of society and in many cases not permitted to own land. This was a severe limitation in an agricultural society, and it left open only the world of commerce to Jews for their livelihoods. Because they had little scope for their efforts other than business and commerce, Jews tended to be successful. This very success, along with religious ideas noted above, probably caused the historic resentment felt towards Jews in Western society.

Anti-Semitism, always present, rose and fell in its severity, usually having to do with other social stress in society at any given time. Because of rapid population growth in the nineteenth century and because of social dislocation surrounding economic modernization, anti-Semitism reached a new peak at the end of the nineteenth century and early in this century. From Russia to the United States and throughout Europe came stories that told of the difficulty of being a Jew in modern times.

Nazi Germany, then, was not alone in its disdain for Jews. But while it did not invent anti-Semitism, it did more than any nation in history to force an answer to "the Jewish question." Deep in Nazi ideology was a conviction about race that was tied to a perverse reading of Darwinian biology. Since, the Nazis said, the highly evolved "Aryan race" was destined to rule the world, racial impediments should be purged from Germany. This "race purification" was addressed in various laws that from 1933 onwards marginalized Jews in Germany. The ultimate logic of this policy—of exterminating the Jews—had never been imagined before, but to this incredible end the Nazis in Germany moved their ideas after war commenced in 1939. What came to be known as the Holocaust began.

The figure six million, often referred to as the number of the killed in concentration camps is, in fact, incorrect. Actually the figure is nine million: six million Jews, to be sure, but at least three million others (gypsies, Communists, the deformed, homosexuals, Catholic and Lutheran clergy, etc.) who did not "fit" into the scheme of things in the "new order" being established in Germany and the nations conquered by German forces after 1939. But, it is to the six million Jews that we make reference in the Holocaust because it was the determined policy of the German government to exterminate European Jewry, a policy, be it said, in which it was nearly successful.

The facts about the concentration camps—now fully known— were slow to come out for two reasons: In wartime, it was difficult to distinguish rumor from fact; more importantly, perhaps, people everywhere denied the carnage of the camps because it was so incredible. Indeed, even the victims themselves—the Jews of Europe—could scarcely believe that it was really happening. But, after 1945, the facts were real enough to all, and the memory of the Holocaust is, at once, a terror and a burden to the modern imagination.

The preeminent person who has done more than any other to chronicle the Holocaust is Elie Wiesel, whose many books and plays include *The Testament, Dawn, The Accident,* and *The Trial of God.* His autobiographical account, *Night* (1960), is a perennial bestseller and the most terrifying book imaginable. Born into a deeply religious Jewish family in the Transylvania region of Austria-Hungary, Wiesel had hopes of being a scholar of Jewish religious literature, the Talmud. In 1941 he was twelve years of age. Over the next four years, his life was—this is no exaggeration—a descent into hell. At first the Jews of his little town of Sighet refused to believe the stories that filtered in about the concentration camps.

It was the week of the Passover in 1944 that Elie Wiesel's world began to crumble: first a decree that Jews should remain in their homes; then a decree that they should wear a yellow Star of David on their clothing; then a neighborhood ghetto was established; then, on the Saturday before Pentecost, the news of a deportation order was given to the ghetto.

The Jews of Sighet were allowed to take a few valuables with them, along with one suitcase of clothing. But, as Wiesel recounts, the arrival at Auschwitz, the air fouled by the smell of burning flesh, began a new phase in his life. "The cherished objects we had brought with us thus far were left behind in the train, and with them, at last, our illusions." The men were segregated from the women, and so, within a few hours of coming to Auschwitz, Wiesel was, in his own words, "parted from my mother and sister forever." Later he was transferred to Buchenwald, where his father died. Alone in the world, Wiesel emerged from the camps at age seventeen in 1945, somehow having survived two of the most notorious concentration camps. But that first night in Auschwitz had determined the future course of his life.

Never shall I forget that night, the first night in camp, which has turned my life into one long night, seven times cursed and seven times sealed. Never shall I forget that smoke. Never shall I forget the little faces of the children, whose bodies I saw turned into wreaths of smoke beneath a silent blue sky.

Never shall I forget those flames which consumed my faith forever.

Never shall I forget that nocturnal silence which deprived me, for all eternity, of the desire to live. Never shall I forget those moments which murdered my God and my soul and turned my dreams to dust. Never shall I forget these things, even if I am condemned to live as long as God Himself. Never.

In the preface to *Night*, French novelist and journalist François Mauriac writes that the concentration camps had killed the dream. What was the dream? "The dream which Western man conceived in the eighteenth century, whose dawn he thought he saw in 1789, and which, until August 2, 1914, had grown stronger with the progress of enlightenment and the discoveries of science." In short, what died in the camps along with the Jews was "hope," that is, "hope" as conceived for humankind in the post-Enlightenment West.

At the same time as the horrific experiences of Elie Wiesel, events were shaping up that would lead to the small city on the other side of earth: Hiroshima. It was a city largely unknown to prior history in the West, but its name forever after would be remembered as the place where the nuclear age began.

Of the technical advances in war evidenced in the Second World War, all paled in significance to the development of *the* weapon, the atomic bomb. Mostly through American and British scientists, and partly through the work of recent emigres from Hitler's Europe, a team was brought together to work on the "Manhattan Project," the code name given to the bomb project. They worked secretly (Vice President Truman did not know of the project's existence until he assumed the presidency in April 1945) and quickly (fearing that the Germans would beat them). After some tests on the deserts of the American Southwest, the bomb was ready, and Hiroshima was bombed on August 6, 1945, with a similar bombing at Nagasaki three days later. Hiroshima and that first bomb are deeply imprinted in Western memory.

In 1945, Hiroshima was an industrial city of some 270,000 people. The conventional American bombing, already begun on other parts of Japan, had bypassed Hiroshima, though its citizens had been prepared for its beginning. But no arrangements in civil defense could have prepared the population for ensuing events. For days after August 6, the residents of the area could not understand what had happened. There was no explosion that any resident could recall, only a "noiseless flash of light." The bomb apparently exploded above the city center, with a heat of 6000°C. So intense was the heat that clay roof tiles, whose melting point is 1300°C, had dissolved at six hundred yards from the center. With heat like that, there is no need to inquire too deeply into what happened to the patients in Shima hospital, virtually at the center.

Telling the story of Hiroshima is difficult precisely because there is no other story quite like it and because the participants themselves had little ability to relate the events of August 6 to the "real" past or future. Yet, in John Hersey's *Hiroshima* we have a vivid and comprehensive account of the story because he brings it down to human scale. The book starts at 8:15 A.M. on August 6— the moment the bomb went off—and it follows the lives of six survivors. There is no other way for us to comprehend the meaning of Hiroshima (knowing that, of the 270,000 residents, 100,000 died does not really help us). There were real people, with real names, who survived. Through the eyes of these six we know the story: Dr. Masakazu Fujii, Dr. Terufumi Sasaki, Mrs. Hatsayo Nakamura, Miss Toshiko Sasaki (no relation to the doctor), Father Wilhelm Kleinsorge (a Jesuit missionary from Germany), and Rev. Kiyoshi Tanimoto (a Methodist minister in Hiroshima). They represent the *hibakusha*, the ones who survived the bomb. Their lives, illnesses, rejections, and fears, as well as their triumphs, are the stuff of Hersey's book.

Nearly twenty years after the bomb, Rev. Kiyoshi Tanimoto was asked to write an editorial in *Chugoku Shimbun*, one of Hiroshima's evening newspapers, about the remembrance of August 6 and about the memorial erected in Hiroshima.

The sentence inscribed on the memorial Cenotaph—"Rest in peace, for the mistake shall not be repeated"—embodies the passionate hope of the human race. The appeal of Hiroshima . . . has nothing to do with politics. When foreigners come to Hiroshima, you often hear them say, "The politicians of the world should come to Hiroshima and contemplate the world's political problems on their knees before this Cenotaph."

President Franklin D. Roosevelt said of the bombing of Pearl Harbor by the Japanese air force on December 7, 1941, that the date "will live in infamy." Surely August 6, 1945—when the age of nuclear terror began—will also live in infamy.

Conclusion: Sorting Out the Postwar World

The events of the twentieth century were as unwelcome as they were pervasive in their impact. The rose-colored glasses of the nineteenth century had—in the view of many—been smashed. But what was the new "reality" to look like? Here is a major point: The intellectual leaders of our time have found no pattern of agreement on the nature of reality. All that we are certain of is that—somewhere, somehow—in the midst of the events of the twentieth century, modern culture (Enlightenment hope) "died."

A glimpse at the artifacts of contemporary culture will demonstrate this. After Picasso, painting and sculpture went "beyond" representationalism. Picasso's successors thought art should not so much represent life as reflect the artist's feelings about life. After Henry James, a novel would not merely tell a story but probe the psychology of the characters. After Pound and Eliot, poems would have no traditional structure and coherence and surely would not rhyme. After Brecht, the theater would not so much entertain patrons as challenge them to ask essential questions about reality. After Bonhoeffer, theology would seek to discuss God outside the patterns of religious language that had "died" with the culture that gave it.

Other names could be given—Beckett, Sartre, Camus, Bergman, Pollack, Cage—but the point would be the same. The "message" of the cultural elite in postwar Europe was one of disconnectedness and despair. (Readers wanting to pursue this theme more fully might consult a Christian analysis by Hans Rookmaaker, *Modern Art and The Death of Culture*.) At first the work of a cultural elite, the message of brokenness and disconnectedness filtered down into popular culture as well. "Rock 'n' roll" as a musical form begat a "rock culture," which with its own language, uniforms, and rhythms has become an international youth culture. Whether in Berlin or Boston, London or Los Angeles, Moscow or Memphis, musical artists such as the Beatles, Mick Jagger, Rod Stewart, David Lee Roth, and Madonna give popular voice to the essential incoherence of life, and, oftentimes, the desire for personal gratification because that is all there is.

The above is not meant to endorse such a view of life but to see its origins and to understand its intent. Moreover, it is a message everyone knows, that Western society has been pummelled by events in this century that have called its most basic assumptions into question. If humankind is not rational and progressive, then what hope is there? If events have shown us humanity (not really inhumanity), how can we believe in a future together on this small planet? If there is no future and no past, we might as well live for an eternal present in the company of "Lucy in the Sky with Diamonds." To a worldview that had promised "satisfaction guaranteed," Mick Jagger laments with anger that he "can't get no satisfaction." Alike does Samuel Beckett see the futility of "waiting for Godot" that Simon and Garfunkel see in America turning its hungry eyes in search of Joe DiMaggio. In short, a generation arose in this century for whom, as Gertrude Stein said, "all gods were dead, all faiths shaken."

It seems clear that a naive faith in humankind's ability to solve its own problems was severely shaken by the events of the twentieth century. But the "shaking of the foundations" was more deeply felt in Europe than in North America. It is to America that we now turn to see if anything of the Enlightenment hope can be salvaged for our time.

Chapter 16

THE AMERICAN COUNTER-CHALLENGE: AN ATTEMPT AT MODERNITY REAFFIRMED

The circumstances in Europe after 1945—deeply and desperately felt by most Europeans—were at marked contrast to North America, especially the United States. In Europe, victors and vanquished alike were exhausted. There did not seem to be the will to reassert the sort of confidence and optimism that had been typical in the West since at least the eighteenth century. No one in Europe would dare to invoke Leibnitz, that "this was the best of all possible worlds" with all things "growing better and better every day." While Europeans may not have been ready to throw in the towel entirely, neither were they prepared to listen to naively innocent talk about progress and the future. In the United States, and similarly Canada, there was a discernibly different spirit, born of different experiences. In America after 1945 there was a sense of confidence and optimism that was a reaffirmation of historic Western ideas about progress. In the postwar era, America became the new proving ground for the Enlightenment and its faith.

America and the Future

It was not that Americans were "naive" and Europeans "realistic" but that the events that had ended "innocence" in Europe were either not experienced directly by Americans or they were not as deeply felt. Americans were not necessarily foolish to try to reaffirm the doctrine of progress in the second half of the twentieth century. The conditions that once had generated faith in the future for all Western people appeared still to be true for America: material abundance, political and religious freedom, and technological advance. For many, if not most, Americans, it seemed that if Europeans could not reaffirm faith in progress—and in humankind's ability to solve its problems—it was largely a failure of nerve. For America the postwar era was to be a testing time. What was being tested was whether or not a people could bring into reality the whole vision of a liberal society; in short, whether or not a free people could survive and thrive and fulfill the dreams for humankind first envisioned in the eighteenth century.

The notion of a successful liberal society—and of America proving the worth of it—accorded well with American self-definition. As discussed in an earlier chapter, America's "historic role" in Western civilization was—as symbol and reality—to be a figure of renewal. It was an "open land," it was said, where Europe's "huddled masses, yearning to breathe free" could come. It was, moreover, a society "open" in many ways. People could come here and, notwithstanding race, religion, or social status, rise as far as their abilities would take them. This was the essence of "the American Dream," that in this free and open place, people could take life in both hands and make of it what they chose. As discussed earlier, America was, in a certain sense, an invention of European wishful thinking in that in America Europe might be "born again."

As historian-economist Robert Heilbroner (1961) has written,

At bottom, a philosophy of optimism is an historic attitude toward the future, an attitude based on the tacit premise that the future will accommodate the striving which we bring to it. Optimism is grounded in the faith that the historic environment, as it comes into being, will prove to be benign and congenial, or at least neutral to our private efforts.

Being an American, then, carries with it an attitude about the past and the future. The American ideology is a historically progressive one, believing in an ever-expanding liberty and an ever-increasing standard of life. But even more than that, it infers that "the past" has a "future." Or, put another way, there is an unbroken line of development from the past, through the present, to the future.

America—the land of the future—is, in a fundamental sense, a place where the ideology of the Enlightenment was to be tested. The idea of time in Western thinking before the Enlightenment was specifically oriented toward the secular future. It was only after the advent of modern thought that people in the West began to think about the future in hopeful terms. If there is one thing that differentiates premodern and modern thinking, it is the historic attitude towards the future. The United States, as nation-state, is the only Western nation whose life span is identical to the life span of the modern, Enlightenment idea of progress. The catastrophic events of Europe in this century shattered a belief in progress and in optimism about the future. An important distinction needs to be made here: It is not that individuals in Europe lack *personal* optimism (about their careers, their families, etc.), but they lack *historic* optimism (about culture and society), or the sense that history is "going somewhere."

White America has never had "a past," that is, a past it wanted to be rid of. America did not have a premodern history that suggested changelessness, nor did it have a "peasant" heritage that resisted change. It did not have a noble class that required democratization, and its established churches in a few colonies were soon disestablished. As the early twentieth-century thinker William James said, America was an attempt to do without history, i.e., the burdens of history. An essential part of distinguishing Europe from America was that Europe was burdened by a past. If America was to take notice of the European past at all, it was as a negative example, not to be emulated but overcome.

Americans had few doubts about the world they were creating in the nineteenth and twentieth centuries because of a prior, if tacit, acceptance of the benign nature of historic change. Americans looked on the effects of technology, on popular political institutions and dynamic economic institutions, and they were optimistic. What Americans *meant* by optimism is that they saw no real conflict between their national goals and the flow of Western history. Americans believed themselves to be masters of their own fate because, to a remarkable degree, they were. Physical isolation from the troubles of Europe, and weak neighbors who could not threaten the national will, gave Americans a belief in the inevitability of progress. Furthermore, they possessed all the natural resources for industrialization, as well as broad and fertile lands that would yield foodstuffs in abundance. When H. G. Wells visited America early in this century, he wrote a little book about his impressions, *The Future in America* (1906). He noted correctly that Americans possessed a kind of "optimistic fatalism" that whatever the future would hold for America would be good and should be welcomed.

Economics and Ideology

Economic performance, more than anything else, gave Americans confidence. As they looked back at what had happened in the nineteenth century, they, and Europeans, too, could not fail to be impressed. In 1800, America was a small nation along the eastern seaboard, largely agricultural and rural, possessing a population of five million, twenty percent of whom were slaves. By 1900, it was a far-flung continental domain, strong in industry and agriculture, with a dynamic population of over 100 million. America had become the world's leading economic power, a position it has maintained for most of this century.

The performance of the American economy has been the historic driving force of American optimism. This fact has astonished foreign visitors for a century and a half: how dynamic American society is in its workings, and how the enormous economic momentum is the force behind it all, both for persons and for institutions. If there is a chosen form of American response to what Robert Heilbroner calls "the closing in of history" it is in the realm of economic growth and the consequent rise of the standard of life. Consumer society, then, is a primary part of American ideology. Whether spoken or unspoken, there is a broad and deep belief that economic abundance is the engine that makes America go.

A clear example of the depth of consumer society as ideology was the reaction of Richard Nixon's confrontation with Nikita Khrushchev in Moscow in 1959. Nixon, as vice president, was representing the United States at a world trade fair in Moscow. Khrushchev, as premier of the Soviet Union, was also on the grounds in an official capacity. The two were not scheduled to meet. But when they did meet, they walked around together, debating Russian and American plans for the future. They walked up to a model of an American home, and standing at the side of an up-to-date American kitchen, Nixon remarked that this—the kitchen—demonstrated the superiority of the American system. An economic system that could make modern kitchens widely available, said Nixon, was surely the system of the future. When these words—along with a picture of Nixon and Khrushchev standing near a washer and dryer—were relayed to the American people, there was broad support for Nixon, not only for "standing up to the Russians" but for stating so clearly "what America was all about." The fact that the substantiation of "the American way" was frankly materialistic did not seem to trouble most Americans. While it was understood that Americans had spiritual values, it was nevertheless economic dynamism that would ensure the nation's future.

In this book we have tried to avoid simple moralisms, and we will not abandon that policy here. It would be simplistic to accuse the Americans—as some have done—of being a materialistic nation and, in that respect, no different from the Soviets whom they oppose. While there are unlovely aspects to American society—now and in the past—it is nevertheless largely true that Americans have been unembarrassed about their riches. Moreover, while there has been a maldistribution of wealth, it is nevertheless true that the dynamic American economy has brought an unprecedented range of goods and services into the possession of ordinary people. For most Americans, most of the time (slaves and native Americans consistently excepted), America has made good on the promise of the American Dream, at least relative to other peoples in other nations. For most Americans—whether rightly or wrongly—there was no real incompatibility between moral success and economic success. And, if there were calls for reform in the past two generations (for example, from black Americans), those calls did not question basic American assumptions but asked that all citizens be brought "into the mainstream" of American life. The American philosophy, then, was one of expectation. A person could have optimism because the society was optimistic. Moreover, there seemed to be good grounds for blending personal and historic optimism because the evidence seemed indisputably to be there in the great socioeconomic success of America.

As the twentieth century draws to its close, is there still an adequate base for this historic hope? If there is hope for humankind in the West, is it still in America that it will be proven or disproven? In short, is America "Atlas," who must hold the world on its shoulders? And, if America should prove unequal to the task, does this necessarily mean that the hope—born in the eighteenth century—is no longer to be held out for all humankind? It is to these sorts of questions that we must now turn.

The Challenge of Events

Events caused Americans—like Europeans in the prior generation—to begin to question both their historic role of renewal for the West and their own belief in the inevitability of progress. Indeed, by the late 1980s, a consensus had emerged across the ideological spectrum that "something" had gone wrong. Writers on the left (Christopher Lasch), the right, especially the Christian right (Francis Schaeffer), and from the academic center (Robert Bellah) agreed that America was adrift. There seemed to be no clear pattern to events that people would call progressive, and there seemed to be a great deal that threatened stability. Moreover, America had seen itself as a nation of "doers," but now other forces impinged and threatened independence of action. Now they were being done-to rather than doing. "History"—once seen to be the thing *they* would make—was no longer the benign force to be welcomed but a malign phenomenon that exacted much and threatened worse.

Robert L. Heilbroner, in his best-selling book *An Inquiry into the Human Prospect* (1974), helps in organizing some current attitudes in America. If there is a new mood in America—reversing historic optimism—it is well for us to look into the origins of malaise in the American spirit.

The first level of confidence-shaking events is what Heilbroner calls *topical*. By topical we do not necessarily mean "superficial" but rather the sorts of events that have come into daily discourse among Americans since, say, 1960. Foremost among these sorts of events was the Vietnam War. The meaning of Vietnam is still being developed and sensed, but we can at least say that it threatened every aspect of American ideology: America's belief in the invincibility of its military power (Could anyone really believe that America might be defeated by a ragtag guerilla army from Hanoi?); the trust that Americans had in their government (Could anyone really believe that America would fight an "unnecessary" war?); the morality of American actions (Could anyone really believe that American forces would engage in systematic poisoning of the Vietnamese countryside or the massacre of innocent civilians?). By the end of the Vietnam War, many Americans *did* begin to believe the hitherto unbelievable about themselves and their nation. In his excellent book on Vietnam, *Backfire* (1985), Loren Baritz points out that defeat in Vietnam called into question two main American assumptions: its cultural superiority to other, "backward" nations and its military-management procedures. Americans saw themselves as a people with a superior cultural ethos and with "standard operating procedures" that would lead them, of course, to victory. Given such prior self-understandings, defeat was a bitter pill to swallow.

Another sort of topical event that caused American confidence to be shaken was the assassination of President Kennedy. Even though Americans knew that political leaders had been shot at many times throughout the nation's history, and three presidents murdered, we thought of those acts as aberrational—until November 22, 1963. John Fitzgerald Kennedy, president of the United States, had gone to Dallas, Texas, to help heal a political rift in the Texas Democratic party. Driving through Dallas in a motorcade, the president and his wife were in an open car. A lone gunman crouched in the Texas School Book Depository—a building along the route—and shortly after noon, he shot John Kennedy. Later, at Parkland Hospital, the president was pronounced dead. Frank McGee was anchoring the NBC news desk that day, and, in another first, this veteran newscaster wept openly on live television—a sign for the emotions of all viewers.

The funeral on the next Monday—November 25, 1963—was one of the main media events of our time. The requiem mass was to be sung at St. Matthew's Cathedral in Washington, D.C. The processional left the White House at 11:30 A.M. The casket was drawn through Washington—the bagpipers of the British Army's "Black Watch" regiment played for the march, at Queen Elizabeth's command—and drew up to St. Matthew's at precisely 12:14 P.M. The nation and the world was aware of the time. Chicago's Loop was deserted. In New York's Times Square, worldly cabbies stood in silence outside their taxis while two Boy Scouts played taps in front of the Astor Hotel. The nation's transportation system was frozen: Trains did not leave, and those in transit had been stopped in cities, woods, or on mountain trestles; Greyhound buses pulled off the road; airplanes awaiting takeoff cut their engines; the roar of subway systems in Boston and New York was stilled. The world took note as well: At that precise moment, the Panama Canal was closed; police in Athens stopped evening rush-hour traffic; in Berlin, a crowd gathered where Kennedy had given his

"I am a Berliner" speech; in Vietnam, where there were a few American "advisors," troops fired salutes. Americans, especially younger ones, remembered Kennedy's inaugural address, given about a thousand days before: "Let the word go forth from this time and place, to friend and foe alike, that the torch has been passed to a new generation of Americans." President Kennedy's life-affirming zest, his brilliance and wit had seemed to be a new style for America—unafraid of ideas, change, and the future. With Wordsworth, many people of Kennedy's "new generation" could say, "In that dawn 'twas bliss to be alive, but to be young was very heaven." With Kennedy's death, while they realized that life must go on, they had nevertheless lost some of the best of themselves. They might live and love and laugh, but they never would be young again.

The murder of John Kennedy—stunning enough in itself—was to be but the beginning of the political violence of the 1960s. In 1968, the American people, already being torn by the Vietnam War, were pummelled still further by the murders of Martin Luther King and Robert Francis Kennedy. The news from Memphis and Los Angeles, of gunmen ready to cut down the nation's leaders, was profoundly disturbing to most thoughtful Americans. This disturbance gave rise to further questions about America's future, in light of the violence seemingly sweeping the nation.

The eruption of racial violence in the major cities provided additional evidence that everything was not well in the land of the free. In the first two-thirds of the twentieth century, a vast migration had gone on, which saw the movement of millions of black Americans from the rural South to the great cities of the North and the West. Within a few years, the rage of blacks who felt oppressed erupted in New York, Newark, Detroit, and Los Angeles. President Lyndon Johnson appointed a national commission to inquire into the causes of such massive civil disorders. The commission, chaired by Illinois governor Otto Kerner, issued its report in 1968. Its most famous remarks were as follows: "This is our basic conclusion: our nation is moving toward two societies, one black, one white—separate and unequal. . . . Discrimination and segregation have long permeated American life; they now threaten the future of every American." After a generation of apparent success in civil rights reform, this was another bitter realization that hard times lay ahead for Americans.

As well as these topical concerns—coming from events that directly intruded into American lives—there were *attitudinal* changes that, though not as marked and obvious as those discussed above, nevertheless were important in reinforcing and underlying the sense of unease about the future. In general, what is meant here is the sense of assurance that we have control over events. More specifically, this came to the fore in those sorts of areas where a pragmatic, managerial approach had been hailed as the solution to problems. Economists talked of "fine tuning" the economy, and social planners spoke confidently of development. All of this would be a natural result of "growth," which, if "managed" properly, could bring social success. We now know—although some people are only beginning to know—that such "structural rationality" has definite limits and that infatuation with "technique" can have disastrous and dehumanizing consequences. This attitude had arisen out of what the French theologian Jacques Ellul has called the "technological mentality," i.e., that if a problem exists, technology will find a solution; if there is a question, technology will find an answer in the general context of "growth."

The first book to call attention to the unintended and unacceptable consequences of growth was Rachel Carson's *The Silent Spring* (when no robins would announce spring because their ecological support system had collapsed). Whether one stands in Riverside, California, in midsummer, with eyes smarting from mustard-colored air, unable to see the San Bernardino mountains, or in Bangor, Maine, with woods thinning because of "acid rain," or in scores of communities where toxic waste dumps have fouled the water supply, one knows that environmental collapse is all too possible if unrestricted growth continues. Whether in major cities that seem ungovernable or in rural areas where traditional family farming seems unsupportable, there is a growing suspicion that "things" are not really subject to our management or control and, indeed, that there are side effects to economic growth that are as bad, or worse, than the problems that growth was supposed to overcome.

This matter of technique unrelated to the realities of the human condition had turned on a kind of *technological imperative*, i.e., that if you *could* do something you *should* do it, because progress was always better. In medicine, for example, it was always believed that newer and better techniques would *always* be better for humankind. In many cases, of course, that is so. But techniques shorn from moral consequences can—and do—lead to unacceptable ends. Abortion—the termination of unborn life—had once been a risky operation but now is a safe and easy surgical procedure. Always legal throughout the first half of American history, it was criminalized around 1900 in most states and then decriminalized by the now famous Supreme Court decision in 1973, *Roe v. Wade*. Neither the courts nor most doctors performing abortions will enter into discussion about the meaning of the latter's actions. It is a surgical procedure that is safe and available, and like any other medical service, it is said, people must be allowed to be "free to choose" it if they wish. To many Americans in the Judeo-Christian tradition, this represents an outrageous state of affairs, but official doctrine in managerial America has it that medicine is morally neutral and is the purely private choice of patient and doctor. To many Americans, this is another index of the way "things" have gotten out of control.

Beyond topical and attitudinal changes, there is another realm of revision in expectations that is vital to the unease about the future we have been discussing. We call this change *civilizational* because it is larger than one nation and not the fault or the problem of one nation. This point allows us to reconnect our discussion to the larger discussion about Western civilization. This realization is as vital to our argument (in this book) as it is difficult for Western people to accept. The main thrust of Western civilization—and in America this was to be tested—was to promise people that material improvement would satisfy the human spirit. Whether on one side of the Iron Curtain or the other, all economic and social systems since the Enlightenment are frankly materialistic, i.e., they believe that material improvement will result in spiritual satisfaction.

A Christian Response to Events

It would be easy for Christians to become moralistic at this point and quote Scripture ("What shall it profit a man . . . ?"), but it is better for us to stick to analysis. Yet, this theme of not being "satisfied" with material success is a major refrain in the disquiet in America and other nations of the industrial West. In one of the most significant books of the 1980s, *Habits of the Heart* (1985), Robert Bellah and his associates survey contemporary America and find it lacking. Using both historical and sociological techniques, the authors do not discuss the people who have lost out in the economic struggles but those who have won. The "winners" who are the subject of the book seem, on the surface, to validate the American Dream. They work hard, and, with a little bit of luck, they have succeeded. Yet they are unfulfilled, and deep within them is a sense of longing for something more. But, more tragic still, these winners do not even have the language to express themselves. As Bellah says, they speak the language of the prevailing American ideology—individualism—but as successful individuals they know, down deep in the unspoken realms of their being, that their success has not brought the satisfaction they sought. The book mentions a "second language" that Americans used to know but have largely forgotten. It is a language rooted in the biblical understanding of humankind, in which connectedness, belonging, solidarity, and mutuality are the hallmarks. A biblical understanding of the self-in-society is one that does not celebrate individualism but a corporate quality that sees persons as members one of another. It is only when people recover their places in "communities of memory," says Bellah, that the satisfaction people so desperately seek will be found.

Readers of this book will already have noted that the conclusion to this chapter might have been predicted, given our previous analysis of the Enlightenment. Since the rationalism of the Enlightenment is incompatible with Christian belief, and since America was to be a testing ground for the progressive beliefs of the Enlightenment, it was always clear—on our terms—that such a test would be a failure. As in Europe, so in America: Events crashed through the dreams and visions of progressive hope that the dominant worldview of the Enlightenment was threatened and overtaken. Lest readers think I have overstated the case here, let us hear again just what that American version of Enlightenment progressivism sounds like. The following is a piece written in 1965 by Clark Clifford, then secretary of defense in the Johnson administration.

WHAT IS AMERICA?

What really is this country that brave men, hope blazing high in their hearts, once called the New World?

Is it merely a geographical location, defined by latitude and longitude on a chart? Or is America not more profoundly an idea—an affirmation, defined politically by a principle and a philosophy that have fired men's aspirations around the globe for nearly two centuries.

Perhaps America might be described as a dividing line in the ancient argument about man and his purposes.

This Nation was forged in a furnace of faith—a faith that free men would prevail no matter what the struggle.

The Nation's fiber was strengthened and tempered by the battle against those who have tried to impose limits on the Nation's belief in itself.

This Nation has found power in welding its people together in a common dedication—not to a dreary uniformity but to a daring diversity.

If this Nation is characterized by any single and unique quality out of the restless welter of opinion that a devotion to democracy demands, it is the stubborn belief that progress is our destiny—both individually, and as a society—and that no barrier to that destiny can be built that a determined America will not breach.

This Nation has never had much time for the past and is forever impatient with present. From the very beginning, our chosen time-frame was the future. Our motivating force has been to fashion a greater prospect, not only for America, but for free men everywhere. We have faced fearful problems in the past and have solved them. We will meet those of today and surmount them.

As for tomorrow, I can promise only new and even more complex trails in the glorious and ever ascending journey on the path to greater human progress. For those to whom much is given, much is expected.

How much, and to what extent, readers of this book affirm or deny the above on Christian grounds is the measure of how deeply they see events having altered America's "historic role" of renewal in the Western world.

Chapter 17

COMING TO TERMS:
THE "CRISIS OF THE WEST"
AND CHRISTIAN HOPE

As we bring this book to a close, there is a temptation that needs to be resisted: to become didactic and moralistic and to say what we "should" do. History—as we have said several times—is not apologetics, and to shift our focus in conclusion would be inappropriate. The historical task, after all, does not involve "setting things straight." Rather, historical study is about understanding other humans in the past, and in that understanding we may understand ourselves better. At the same time, this book does have an ideological timbre, seeking to integrate Christian commitment and historical study. We have tried to engage the same subject matter that any honest student would study, but we have done so out of the concerns we have called "a kingdom vision."

Our Place in the Dialogue

Part of the historical dialogue is the recognition that the presupposition of the historian matters very much in the way history is done. This is true for historians of all religious commitments or of none. So, part of the historical dialogue is becoming self-conscious of ourselves as historical beings and of our place in the dialogue. As we said in the first chapter, Christians have a "whole" view of life that takes into account the three-dimensionality of human existence—time, space, and spirit. We insist that any account of human life that does not take all three seriously is an unnecessarily limited account. Not only is it limited, we insist, it is fundamentally flawed. Yet, it will do no good for us to rail against other historians who try to write some version of "scientific" (neutral) history. They who either cannot or will not see what we see will not be argued into seeing things our way. Recall the gestalt picture in the first chapter: If people cannot or will not see that both images are "really" there, shouting at them and arguing with them will not bring the desired result. But, if we point out—with calmness and graciousness—that the nose of one is the chin of the other, people will perhaps say, "Oh, I see."

The writing and reading of this book is like that. In answer to the question, What does it mean for a Christian to study history? this book replies, "Would you let me tell you how one might view the history of the West?" Without seeking a mean-spirited confrontation with our secular colleagues, we tried to say that—in our self-awareness—our placement in the kingdom of God was vital to our "worldview," i.e., how we look at all of life. To us, religion is not an aspect of life, but, rather, religion *is* life. And all human actions are variations on a theme: God has acted, and we react, either positively or negatively. However, we were careful to note the ambiguities and ironies that prevent us from making human experience a simple black-and-white exercise. The problem remains, however, that others without our faith commitment do not always see history (enter into the dialogue with the past) the way we do. Perhaps an extended example will help. I will draw upon, and adapt for our purposes, the insight offered by Alisdair MacIntyre, in his much-acclaimed book *After Virtue* (1981, revised 1984).

The Failure of the Enlightenment

A catastrophe has befallen humankind. It has lost "the sense" for the creative and sustaining work of God. And—importantly—

humans no longer realize the nature of the catastrophe they have suffered. But, most secular people would object, since Western history is fairly well open to view, and since most historians do not see a catastrophe, how can Christians say that it occurred? That is exactly the point: the catastrophe was of such a kind that it was not, nor has it been, recognized as catastrophic. Secular persons may, in fact, not recognize that the forces of light ("the city of God") have always been in combat with darkness ("the city of man"). Begun in the Renaissance and completed in the Enlightenment was a steady progression of human assertion that increasingly marginalized the spiritual realm of human existence even as it emphasized the capacities and capabilities of humankind. At first, humans were thrilled by this "liberation" from the "order" of a Christian culture. But, in the end, thrill turned to despair as "the Enlightenment project" largely failed, and "disorder" threatened all instructions. Listen to Alisdair MacIntyre:

History by now in our culture means academic history, and academic history is less than two centuries old. Suppose it were the case that the catastrophe of which my hypothesis speaks had occurred before, or largely before, the founding of academic history, so that the moral and other evaluative presuppositions of academic history derived from the forms of the disorder which it brought about. Suppose, that is, that the standpoint of academic history is such that from its value-neutral viewpoint moral disorder must remain largely invisible. . . .

If this were to be so, it would at least explain why what I take to be the real world and its fate has remained unrecognized by the academic curriculum. For the forms of the academic curriculum would turn out to be among the symptoms of the disaster whose occurrence the curriculum does not acknowledge.

The secular-scientific humanism of the Enlightenment (not the Christian humanism of the Renaissance and Reformation) has led humankind down a blind alley. Now "up against the wall," as it were, fully modern people have nowhere to turn, and, more tragically still, they have largely lost the memory of how they got into all of this and what remedies there might be for "the human condition." As French historian Lucien Goldmann writes (1973): "History is irreversible and it seems impossible that Christianity should ever again become the mode in which men really live and think." If people are Christians at all, according to Goldmann, it can only be as "a purely inward, psychological 'private matter,' deprived of all influence on life in society."

From the viewpoint of "the spirit of the modern age" it could seem that the assumptions of this book are quite impossible. The writer and most of the readers are Christians, for whom faith commitment is indeed a personal matter. But Christianity is much more than that; it is a coherent view of the world. The premise of this book—and the premise of Christian perspectives on all academic disciplines—is explicitly confrontational to the spirit of the modern age. Modernity, we insist, is based on a huge pretense, what Dutch philosopher Herman Dooyeweerd (1965) called the pretense of the autonomy of human thought. In our perspective on the past—as in our contemporary lives—we need not fall victim to that pretense because we belong to a community of people who can remember what it was like before the catastrophe. While life and thought were never perfectly Christian—there was no golden age that we would restore—we remember that there is a way to bridge the gap between the "real" and the "ideal." We understand that there always was a tension between "the already and the not-yet" of the kingdom of God that has come and is yet to come. We also understand that the two kingdoms will not always be present but that at the end of what we call "time" there will be a consummation. In philosopher Nicholas Wolterstorff's words, justice and peace will one day embrace in the kingdom of God's *shalom.*

This concluding section of the book is called "Coming to Terms," because we cannot go beyond the present. The canon of human history is not yet closed, so we cannot make grand conclusions. But we can at least come to terms with the human story—our story—in Western civilization. It is a story of both triumph and tragedy, of ecstasy and agony, of hope and despair. In its large contours we can begin to see some patterns. At the end of the twentieth century the Western story presents a rather bleak picture for Christians.

Some Christians would prefer to ponder the prophecy sections of the Bible and use Daniel and Revelation to interpret "the signs of the times." For historical students—having noted the ebb and flow of history—apocalyptic ponderings seem a rather sterile exercise. There is greater insight, and comfort, too, in recalling the great work of St. Augustine, whose writings have animated Christians through the ages. When Roman authority was crumbling and the world as the people of classical times knew it was passing away, Augustine gave a reminder to Roman (and later) Christians. He pointed out that earthly cities pass away. Whether Rome in his time, England in the nineteenth century, or America in the twentieth century, Christians do well to avoid giving too much allegiance to regimes and nations that will surely pass away. Christian hope, in short, is as viable as ever, once we come to terms with the spirit of our times and idols it has created.

EPILOGUE:
GLIMPSES OF CHRISTIAN HOPE

Coming realistically to terms with life at the end of the twentieth century leaves some people with a sense of doom and gloom. Because Protestant Christianity in North America has long been associated with nationalist and progressivist ideas, giving them up is difficult. To say that one is loyal to the kingdom, not to the nation, is all very well—and it needs to be said by Christians—but it nevertheless leaves some people with a kind of letdown. For some Christians it is too difficult a world to live in if America is not the focus of Christian attention and if America is not leading the world into a progressive future. The question is, in short, if we have to give up *all that,* then what can history tell us about Christian hope? And, the questions persist, what can we *do* to be a kingdom servant?

The last part of the questions should be dealt with first. There is too much emphasis in our culture on doing rather than knowing. Let us be liberated from the compulsion to *do* something. Sometimes it is better to rest in *understanding,* rather than always moving (sometimes too) quickly to *doing* something about it. At the same time, historical understanding does not leave people helpless in the face of needing to act. To the following questions we would turn in this epilogue: Are there any examples in history of people who served the kingdom, and how can we connect with their stories?

The following five stories are appended to this book in the belief that they will give glimpses of Christian hope. In all cases they show that discipleship is costly. The persons to whom we refer in these vignettes are well known to certain expressions of Christianity and largely unknown to others, although Mother Teresa is probably known to all. Anabaptists will know of Dirk Willems, but perhaps he is not well known among Methodists, whose Francis Asbury, in turn, may not be known well elsewhere. The evangelical and the Free Church/Baptist expression may know William Wilberforce quite well. But Reformed people may not have heard of him, while being perplexed that other Christians know little of Abraham Kuyper. But, as we will see, these stories belong to all Christians for they disclose the testimony of faithfulness to the kingdom. These are stories for all of us to share because they are a part of the "cloud of witnesses" that encourage the whole church to be faithful.

Dirk Willems (d. 1569) was a humble person, otherwise unknown to human history. Yet in his prosaic story we learn a sign of kingdom faithfulness. In the Netherlands in the mid-sixteenth century, followers of Menno Simons (Mennonites) were not allowed to practice their faith openly. The Mennonites were outlawed, and many of the adults baptized by Simons were executed. The secret assemblies of the Mennonites were referred to as "conventicles" by the authorities and were illegal. At such assemblies, there was preaching, teaching, and "believer's baptism," which gave the Mennonites, and others, their name because they were allegedly "rebaptizing" already baptized persons.

One day, Dirk Willems was seen by a member of the police who were out catching thieves and Anabaptists (both thought to be threats to civil authority). Willems's ideology did not allow him to resist the authorities, but he could try to elude them. In trying to get away from the catchers, Willems crossed over a partly frozen pond. His pursuer, however, fell through the ice and was in danger of drowning in the icy water. His cries for help went unheard except by Dirk Willems. What should he do? If he kept running his life and ministry would be saved. But he could not block out the calls for help by a fellow human being. So, he returned to the icy scene and pulled the would-be catcher to safety. The latter was so grateful to be saved, and astonished by Willems's charity, that he was inclined to let Dirk go. But higher authorities reminded the catcher of his oath to serve the civil government. Willems was indeed arrested and, a few days later, was executed by being burned at the stake, in the town of Asperen. He was faithful to the cause of the gospel, as he saw it, and to the cause of human solidarity. He was faithful to death. He did not accomplish much during his life, at least by worldly standards, but he was faithful. For Dirk Willems, and for Anabaptists the world over, that is more than enough to say.

Mother Teresa (b. 1910) is the most famous Roman Catholic nun in the contemporary world. Born in Skopje, in what is now Yugoslavia, her original name was Agnes Bojaxhiu. At the age of twelve she sensed her calling to be a missionary to the poor. Her interest in India was focused early in her life through reports from Jesuit missionaries in Bengal. In 1928 she left home to join the Sisters of Loretto, an Irish community with a mission in Calcutta. She trained at Loretto institutions in Dublin and Darjeeling and took her final vows as a nun in 1937.

It was during her time as a teacher at St. Mary's High School in Calcutta that she received a more specific call to service (in her words, "a call within a call"). Her heart was struck by the suffering outside the school's walls: the homeless, the lepers, and the destitute. Above all, she became concerned for those dying alone in the streets, without the merest human dignity. She said, "I had to leave the convent and help the poor, while living among them." She left the Loretto Sisters and began her ministry under the pastoral care of the archbishop of Calcutta. In 1950, many young women had joined her and the work was officially organized as the Missionaries of Charity.

The sisters of the community take the traditional vows of poverty, chastity, and obedience. But, unique to their community, they take a fourth vow: "to give whole-hearted free service to the poorest of the poor, to Christ in his distressing disguise." In speaking to journalist Malcolm Muggeridge about the meaning of this vow, Mother Teresa explained, "This vow means that we cannot work for the rich; neither can we accept any money for what we do. Ours is to be a free service, and to the poor."

In 1952 Mother Teresa opened the Nirmal Hriday ("Pure Heart") Home for Dying Destitutes in a building donated by the city of Calcutta. In the words of one of the Sisters of Charity, "It is a shelter where the dying poor may die in dignity." In a world tragically accustomed to poverty, hunger, and large-scale suffering, Mother Teresa believes in her work and that God has called her to it. While she is aware of the millions dying beyond the scope of her ministry, she lives with an attitude of hope, in which it is a blessing and grace to serve the unlovely and unwanted that God has put across her path.

The indomitable spirit has been characterized (by journalist Polly Toynbee) as lacking "any sense of indignation." Mother Teresa reminded Ms. Toynbee that "in the teachings of Christ, there is no rage or indignation, no burning desire to change the horrifying injustices of a society that allows such poverty." Rather, there is only the injunction to love and to turn the other cheek. When asked about a militant and radical Christian approach to social change, Mother Teresa said, "I am called to help the individual, to love each person, not to deal with institutions." The way of Mother Teresa may not be the calling for all Christians. But, they surely recognize in her work an authentic witness to the kingdom.

William Wilberforce (1759–1828) was born into a prosperous English family and, after attending the best schools and Cambridge University, became a member of Parliament at age twenty-one. Not long after that he became a serious Christian and was minded to leave the corrupt scene of politics and life among the upper classes. The massive revival under the Wesleys had, as yet, little impact beyond the lower classes. It was in the nineteenth century that Methodism's full social impact would be felt. At the urging of his friend John Newton (author of the hymn *Amazing Grace*), Wilberforce decided to remain in Parliament as a servant of God.

In Wilberforce's view, the general moral rot in British high society was as much public as private. Adultery and gambling were common among society's leaders, as were the profits made through the slave trade. For most of the remainder of his life, Wilberforce directed the campaign against slavery and the slave trade. He endured much ridicule and opposition. Lord Melbourne perhaps spoke for many people in power when he said of Wilberforce's desire to articulate Christian convictions in parliamentary policy, "Things have come to a pretty pass when religion is allowed to invade public life." But to Wilberforce and his Christian friends, the desire to own slaves or to profit from the slave trade was not a purely private matter. It was a public matter on which Christians had a right and a responsibility to comment and to advocate change. For over twenty years he fought what appeared a losing cause, but, in the end, victory came as Parliament outlawed the slave trade in 1807. Wilberforce carried on for another twenty-six years until the practice of slavery was abolished throughout the British Empire. The movement launched by Wilberforce and his friends had great impact throughout British life, as witness the evangelical presence in campaigns to improve poor laws and prisons. His Christian stand in public life is a great testimony to the fact that a Christian worldview is never a purely private matter, but it has great social impact.

Francis Asbury (1745–1816) was the most important Methodist leader in American history. The Methodists were vitally important in giving American religion its characteristic stamp of revivalism and evangelism. The work of Francis Asbury in organizing

Methodism early in the nineteenth century was important beyond the success of one denomination; it was important to American religion in general.

Asbury had become a Christian as a teenager in his native England. From 1766 to 1771 he preached throughout England and then volunteered to go out to America. He remained in America from 1771 till his death in 1816. The Methodists were not very numerous at the time of the American Revolution, and their ministry reached its lowest point when, in part through the writings of John Wesley, they were identified with the loyalist cause. In the early 1780s, Asbury was virtually alone in his calling to serve the church in America. He was an organizational genius. He never owned his own home but was welcomed in city and countryside among faithful people. He traveled an estimated 300,000 miles on horseback, organizing Methodist congregations throughout the new nation. He had a knack for recruiting the right sorts of lay people to join him and other leaders in traveling the circuit of preaching assignments. In a nation moving west and growing rapidly, the sort of organization envisioned and developed by Asbury was well suited to succeed in a relatively institutionless society. Not only did Methodism minister to the religious needs of millions of potentially unreachable people, it transformed itself from a denomination distrusted by Americans to one of the most characteristically American denominations. Moreover, the desire for holiness was not solely personal (although holiness does start with a person) but social as well. Many of the social reforms of the nineteenth century, especially abolition and temperance, were begun or energized by and through Methodism. Much of the moral tone in American public and private life is the important legacy of Methodism and its premier organizer, Francis Asbury.

Abraham Kuyper (1837–1920) was born into a clergy home in Maasluis, the Netherlands. His long and varied life in church and society was part of a whole, based on his conviction that all of life is the Lord's and that there was "not a square inch" of the world that did not belong to God. So, for Kuyper, the task for Christians was as much social action as evangelism, the practice of scholarship as the practice of piety, the building of the church as the witnessing to political structures.

He was a minister in the Reformed (national) church in the Netherlands *(Hervormde Kerk),* although in 1886 he was instrumental in a secession that was confessionally more orthodox. He was the founder and editor of a national newspaper, *The Standard (De Standaard),* from 1872 till 1920. He did more than any other person to bring matters of Christian distinctiveness to public awareness. He was a political leader. Resigning his active ministry in 1873, he began a career as a member of Parliament that would be capped by his term as prime minister during 1901–1905. He was an educational leader. He was founder of a fully Christian university, the Free University of Amsterdam, in 1880. At the Free University he articulated doctrines of "sphere sovereignty" and "the antithesis" that marked off Christian scholarship from secular scholarship.

Any one of these careers—clergyman, journalist, politician, educator—would have been enough for an ordinary man. But Kuyper was no ordinary man, despite several breakdowns, and he modeled in his diverse life that Jesus Christ was Lord of *all* of life.

These five stories of the lives of Willems, Teresa, Wilberforce, Asbury, and Kuyper are as different as imaginable. Yet, they are all eloquent testimony to a common theme, i.e., that Christianity, as personal faith and worldview, has something profound to say to, and about, Western civilization. Moreover, the memory of these diverse stories gives Christians in the last part of the twentieth century some cause for hope. The telling and retelling of such stories among us puts us back into what Robert Bellah called "a community of memory." With memories intact and integral to our own stories, we can go forward without fear because we, too, know which kingdom is coming and which one we should seek first.

BIBLIOGRAPHY

Adler, Mortimer. *Six Great Ideas.* London: Macmillan, 1981.

Bailyn, Bernard. *The Ideological Origins of the American Revolution.* Cambridge, MA: Harvard University Press, 1967.

Baritz, Loren. *Backfire.* New York: Ballantine, 1984.

Barrett, William. *Irrational Man.* New York: Doubleday, 1958.

Bebbington, David. *Patterns in History: A Christian View.* Downers Grove, IL: InterVarsity Press, 1979.

Becker, Carl. *Everyman His Own Historian.* Chicago: Quadrangle, 1966.

———. *Heavenly City of Eighteenth-Century Philosophers.* New Haven, CT: Yale University Press, 1932.

Bell, Daniel. *The Coming of Post-Industrial Society.* New York: Basic Books, 1973.

Bellah, Robert, et al. *Habits of the Heart: Individualism and Commitment in American Life.* Berkeley, CA: University of California Press, 1985.

Brinton, Crane. *Ideas and Men.* Englewood Cliffs, NJ: Prentice Hall, 1963.

Butterfield, Herbert. *The Origins of Modern Science.* New York: Macmillan, 1951.

Carr, Edward H. *What is History?* New York: Alfred Knopf, 1964.

Chadwick, Owen. *The Secularization of the European Mind in the Nineteenth Century.* Cambridge, England: Cambridge University Press, 1975.

Clark, Kenneth. *The Romantic Rebellion.* New York: Harper, 1973.

Cochran, Thomas C. *Challenges to American Values.* New York: Oxford University Press, 1985.

Colson, Charles. *Kingdoms in Conflict.* Grand Rapids, MI: Zondervan, 1987.

Dawson, Christopher. *The Dynamics of World History.* London: Sheed and Ward, 1957.

Dijksterhuis, E. J. *The Mechanization of the World Picture.* Oxford: Clarendon Press, 1961.

Dooyeweerd, Herman. *In the Twilight of Western Thought.* Nutley, NJ: Craig Press, 1965.

Erikson, Erik. *Young Man Luther.* New York: Norton, 1958.

Estep, William R. *Renaissance and Reformation.* Grand Rapids, MI: Eerdmans, 1986.

Ferguson, Wallace. *Europe in Transition, 1300–1520.* New York: Harper & Row, 1982.

Frost, Frank. *Greek Society.* Lexington, MA: D.C. Heath, 1987.

Fussell, Paul. *The Great War and Modern Memory.* New York: Oxford University Press, 1975.

Gay, Peter. *The Age of Enlightenment.* New York: Time-Life, 1966.

Gilson, Etienne. *Reason and Revelation in the Middle Ages.* New York: Scribners Sons, 1958.

Gordon, Cyrus. *The Common Background of Greek and Hebrew Civilizations.* New York: Harper & Bros., 1962.

Handlin, Oscar. *The Uprooted.* Boston: Little Brown & Company, 1951.

Harvey, Van A. *The Historian and the Believer.* New York: Macmillan, 1966.

Hayes, Carlton. *A Generation of Materialism.* New York: Harper & Bros., 1941.

Hazard, Paul. *The European Mind, 1680–1715.* New Haven, CT: Yale University Press, 1953.

Heilbroner, Robert. *The Future as History.* New York: Norton, 1961.

———. *An Inquiry into the Human Prospect.* New York: Norton, 1974.

———. *The Making of Economic Society.* New York: Norton, 1980.

Hofstadter, Richard. *The American Political Tradition.* New York: Viking, 1948.

James, William. *Varieties of Religious Experience.* New York: Modern Library, 1936.

Kaufmann, Walter. *From Shakespeare to Existentialism.* New York: Doubleday, 1960.

Klaaren, Eugene M. *The Religious Origins of Modern Science.* Grand Rapids, MI: Eerdmans, 1977.

Kuhn, Thomas S. *The Structure of Scientific Revolutions.* Chicago: University of Chicago Press, 1962.

Lasch, Christopher. *The Culture of Narcissism.* New York: Norton, 1979.

Leuchtenberg, William. *A Troubled Feast*. Boston: Little, Brown & Company, 1983.

Lyon, Bryce. *The Origins of the Middle Ages*. New York: Norton, 1972.

Lyon, David. *Karl Marx: A Christian Appreciation of His Life and Thought*. Tring, England: Lion Press, 1979.

McIntire, C. T. and Ronald A. Wells, eds. *History and Historical Understanding*. Grand Rapids, MI: Eerdmans, 1981.

McLellan, David. *Karl Marx*. New York: Viking, 1975.

MacIntyre, Alisdair. *After Virtue*. Notre Dame, IN: University of Notre Dame Press, 1984.

Mandrou, Robert. *From Humanism to Science*. Atlantic Highlands, NJ: Humanities Press, 1979.

Marsden, George M. "The Spiritual Vision of History" in McIntire and Wells, eds., *History and Historical Understanding*. Grand Rapids, MI: Eerdmans, 1981.

May, Henry F. *The End of American Innocence*. New York: Alfred Knopf, 1959.

Mohr, James C. *Abortion in America*. New York: Oxford University Press, 1978.

Moore, James. *The Post-Darwinian Controversies*. New York: Cambridge University Press, 1979.

Noll, Mark, George Marsden, and Nathan Hatch. *The Search for a Christian America*. Westchester, IL: Crossway, 1984.

Painter, Sidney and Brian Tierney. *Western Europe in the Middle Ages*. New York: Alfred Knopf, 1970.

Palmer, Robert R. *The World of the French Revolution*. New York: Harper, 1972.

Ralph, Philip L. *The Renaissance in Perspective*. New York: St. Matthews, 1973.

Reid, Stanford. "The Problem of the Christian Interpretation of History." *Fides et Historia* 5 (1973): 96–100.

Roberts, Frank C. and George M. Marsden. *A Christian View of History?* Grand Rapids, MI: Eerdmans, 1974.

Roberts, Frank C. *To All Generations: A Study of Church History*. Grand Rapids, MI: Bible Way, 1981.

Rookmaker, Hendrik. *Modern Art and the Death of a Culture*. Downers Grove, IL: InterVarsity Press, 1970.

Rostow, Walt W. *The Stages of Economic Growth*. Cambridge, England: Cambridge University Press, 1964.

Rude, George. *The Eighteenth Century*. New York: The Free Press, 1965.

Schaeffer, Francis. *A Christian Manifesto*. Westchester, IL: Crossway, 1983.

Schlossberg, Herbert and Marvin Olasky. *Turning Point: A Christian World View Declaration.* Westchester, IL: Crossway, 1987.

Southern, R. W. *The Making of the Middle Ages.* New Haven, CT: Yale University Press, 1953.

Stace, W. T. *Religion and the Modern Mind.* Philadelphia: Lippincott, 1952.

Stout, Harry S. *The New England Soul.* New York: Oxford University Press, 1986.

Sweet, William W. *The Story of Religion in America.* New York: Harper, 1950.

Tawney, R. H. *Religion and the Rise of Capitalism.* New York: Harcourt, Brace, 1947.

Thiselton, A. C. *The Two Horizons.* Grand Rapids, MI: Eerdmans, 1981.

Thistlethwaite, Frank. *The Great Experiment.* Cambridge, England: Cambridge University Press, 1961.

Thompson, E. P. *The Making of the English Working Class.* New York: Vintage, 1966.

Tocqueville, Alexis de. *Democracy in America.* New York: Vintage, 1960.

Walsh, Brian and Richard Middleton. *The Transforming Vision: Shaping A Christian World View.* Downers Grove, IL: InterVarsity Press, 1984.

Weber, Max. *The Protestant Ethic and the Spirit of Capitalism.* New York: Scribners, 1930.

Wedgwood, Cicily V. *Oliver Cromwell.* London: Duckworth, 1947.

Wenger, John C. *Even Unto Death: The Heroic Witness of the Sixteenth Century Anabaptists.* Richmond, VA: John Knox Press, 1961.

White, Morton G. *Social Thought in America: The Revolt Against Formalism.* Boston: Beacon, 1957.

Wolterstorff, Nicholas P. *Reason Within the Bounds of Religion.* Grand Rapids, MI: Eerdmans, 1976.

———. *Until Justice and Peace Embrace.* Grand Rapids, MI: Eerdmans, 1983.

Wood, Gordon. *The Creation of the American Republic.* Chapel Hill, NC: University of North Carolina Press, 1969.

Yoder, John H. *The Politics of Jesus.* Grand Rapids, MI: Eerdmans, 1972.

INDEX